MANAGING WATER RESOURCES
PAST AND PRESENT

Managing Water Resources
Past and Present

The Linacre Lectures 2002

Edited by

JULIE TROTTIER

and

PAUL SLACK

OXFORD
UNIVERSITY PRESS

OXFORD
UNIVERSITY PRESS

Great Clarendon Street, Oxford OX2 6DP

Oxford University Press is a department of the University of Oxford.
If furthers the University's objective of excellence in research, scholarship,
and education by publishing worldwide in

Oxford   New York

Auckland   Cape Town   Dar es Salaam   Hong Kong   Karachi   Kuala Lumpur
Madrid   Melbourne   Mexico City   Nairobi   New Delhi   Taipei   Toronto
Shanghai

With offices in

Argentina   Austria   Brazil   Chile   Czech Republic   France   Greece
Guatemala   Hungary   Italy   Japan   South Korea   Poland   Portugal
Singapore   Switzerland   Thailand   Turkey   Ukraine   Vietnam

Oxford is a registered trade mark of Oxford University Press
in the UK and in certain other countries

British Library Cataloguing in Publication Data
Data available

Library of Congress Cataloging-in-Publication Data
Managing water resources past and present
edited by Julie Trottier and Paul Slack.
p. cm. – (The Linacre lectures ; 2002)
Includes bibliographical references and index.
ISBN 0–19–926764–2 (hardback : alk. paper)
1. Water resources development. 2. Water-supply—Management.
I. Trottier, Julie.
II. Slack, Paul. III. Linacre lecture ; 2002.
TC405.M34 2004   333.91—dc22   2004019148

ISBN 0–19–926764–2

Typeset by Footnote Graphics Limited, Warminster, Wilts
Printed by Digital Books Logistics, Peterborough

ACKNOWLEDGEMENTS

THE twelfth series of Linacre Lectures was sponsored and organized by Linacre College and the Environmental Change Institute, University of Oxford, and the College is grateful once again to the Director of the Institute, Professor Briden, and his colleagues for their help and support. We owe thanks also to the Centre for Water Research and St Peter's College, Oxford, who were associated with us in organizing the series. Dr Julie Trottier brought her special expertise and inspiration to the tasks of framing the lectures and editing this volume. From beginning to end, all the practical arrangements were managed with unfailing energy and efficiency by Jo Whitfield, the Principal's secretary. I am indebted to all of them.

P.A.S.
Oxford

CONTENTS

NOTES ON CONTRIBUTORS

Dr Peter Ashton is Principal Scientist and Divisional Fellow in CSIR, Pretoria, and Honorary Professor of Water Resources Management and Member of the Advisory Board for the Centre for International Political Studies at Pretoria University (2000–5). His publications include *South African Approaches to River Water Quality Protection* (1995) and *Southern African Water Conflicts: Are They Inevitable or Preventable?* (2000).

Stefano Burchi is Senior Legal Officer, Development Law Service, Food and Agriculture Organization, United Nations, and the author of *Preparing National Regulations for Water Resources Management—Principles and Practice* (1994, 2003).

Sir Ian Byatt, Senior Associate at Frontier Economics, is the former Director-General of Water Services, the head of OFWAT.

Mike Edmunds is Research Director of the Oxford Centre for Water Research and Visiting Professor in Hydrogeology in the School of Geography and Environment, Oxford University. He is the author of over 200 scientific papers on groundwater and hydrochemistry.

Peregrine Horden is Reader in Medieval History, Royal Holloway, University of London. He is co-author, with Nicholas Purcell, of *The Corrupting Sea: A Study of Mediterranean History* (2000).

Maria Kaika is Lecturer in Human Geography, School of Geography and the Environment, Oxford University, and Fellow of St Edmund Hall. She is the author of *City of Flows: Modernity, Nature and the City* (2004).

Martin Reuss is a Senior Historian, US Army Corps of Engineers Office of History, and the author of several books and articles on water in the United States, including *Designing the Bayous: The Control of Water in the Atchafalaya Basin, 1800–1995* (1998).

Paul Slack is Principal of Linacre College, and Professor of Early Modern Social History, University of Oxford. He is the author of *The Impact of Plague in Tudor and Stuart England* (1985) and co-editor of *The Peopling of Britain: The Shaping of a Human Landscape* (2002).

Julie Trottier is Lecturer in the Politics of Water Development at the University of Newcastle, and Thames Water Senior Research Fellow, Oxford Centre for Water Research, University of Oxford. She is the author of *Hydropolitics in the West Bank and Gaza Strip* (1999).

LIST OF FIGURES

LIST OF TABLES

ABBREVIATIONS

AT	Amsterdam Treaty
BT	British Telecom
CM	European Council of Ministers
CSCs	Customer Service Committees
DETR	Department of the Environment, Transport and the Regions
DoE	Department of the Environment
DRC	Democratic Republic of Congo
DTI	Department of Trade and Industry
DWI	Drinking Water Inspectorate
EA	Environment Agency
ECEWTF	European Commission Environment Water Task Force
ECPA	European Crop Protection Association
EEA	European Environment Agency
EEB	European Environmental Bureau
EIA	Environmental Impact Assessment
EP	European Parliament
ESEM	Earth systems engineering and management
ETC/IW	European Topic Centre on Inland Waters
EU	European Union
EUREAU	European Union of National Associations of Water Suppliers and Waste Water Services
FAO	Food and Agriculture Organization
GSS	Guaranteed Standards Scheme
ICWS	International Centre of Water Studies
IWRM	Integrated water resource management
MEP	Member of European Parliament
NGO	Non-Governmental Organization
NRA	National Rivers Authority
OFWAT	Office of Water Services (UK)
ONCC	OFWAT National Customer Council, now Water Voice
OSPAR	Oslo and Paris Commission
PHS	Priority Hazardous Substances
Quango	Quasi-Non-Governmental Organization
RBO	River Basin Organization
RSPB	Royal Society for the Protection of Birds
WFD	Water Framework Directive
WWF	World Wildlife Fund

Introduction

Julie Trottier

THE need for interdisciplinary research on water is now widely acknowledged. Successful flood management, solutions to water scarcity, and adequate sanitation cannot be achieved otherwise. International organizations and national research councils have been promoting interdisciplinary research on water management and water development for several years. Yet, interdisciplinary research efforts generally fail, whether they are directed at water or at any other subject. Understanding the stumbling blocks that prevented successful interdisciplinarity in the past is therefore important. This is a preliminary step to the construction of interdisciplinary methodologies that will allow water issues to be investigated successfully.

A first stumbling-block arises from the issue of the definitions offered by the various disciplines. What is perceived as an objective problem by an engineer or a natural scientist is often described as a transient social construction by a sociologist or a geographer. Consequently the natural scientist often tends to focus on finding solutions to problems whereas the sociologist wants to move away from impact-oriented modes of understanding.

This difference in issue definition often leads researchers to consider each other's pursuits as irrelevant, a phenomenon that constitutes an important obstacle in the quest for interdisciplinarity. How can we build an interdisciplinary theoretical framework together when people are not asking the same questions? Many an interdisciplinary endeavour has shattered at this point and turned into a multidisciplinary project, harbouring parallel researches that defined issues and framed problems differently and each could not possibly feed the other.

If we are successfully to take up the challenge of interdisciplinarity we must start with an understanding of the various definitions and the perceptions of problems generated by the paradigms prevalent in every discipline. This volume of Linacre Lectures allows us to embark on such a path. Every contributor offers an analysis of a water issue as it emerges from his or her own disciplinary framework. This allows us to understand how issues appear differently within these disciplines. Identifying which issues are deemed to be independent and which dependent variables in each discipline allows us to begin a dialogue that can lead to the development of a truly interdisciplinary research framework.

Engineering, as a discipline, has traditionally treated what it defined as 'natural' and 'human' systems as predictable, controllable, and external to the engineer and his actions. The recent emergence of earth systems engineering and management (ESEM) might constitute, at last, an integration of postmodernist criticism within the discipline. '[I]t is apparent that such complex systems cannot be "controlled" in any usual sense: accordingly, ESEM is a design and engineering activity predicated on continued learning and dialogue with the systems of which the engineer is an integral part' (Allenby 2002: 8).

Yet the old belief that water systems, both human and natural, can be and should be controlled still lives on and surfaces in several contributions even within this volume. Mike Edmunds, for example, offers a hydrogeologist's understanding of aquifer recharge. He deplores the fact that policy-makers have been deaf to such results and are proceeding to overextraction of many aquifers with ensuing disastrous consequences. The concept of water governance, rather than water management, offers a route to integrate the recognition that water systems cannot be controlled by one external, impartial manager. The concept of governance recognizes the multiplicity of centres exercising power over water. This is developed further by Maria Kaika in this volume, within a political geography analysis.

This collective volume is therefore an attempt at tackling the vast topic of water management in a multidisciplinary manner. It does not pretend to interdisciplinarity, but should, however, allow us to begin to develop an interdisciplinary research framework since the basic challenge to interdisciplinarity lies in the common construction of a theoretical framework weaving together the contributions of various disciplines to a specific object of study. Here, the contributors have articulated their perceptions of crucial water management issues each within the perspective of her or his discipline.

Mike Edmunds details how hydrogeology allows us to understand the recharge of aquifers. The present rates of extraction often appear totally unsustainable since, as he demonstrates, water is often mined at a much faster pace than it replenishes an aquifer. His vision is clear: hydrologists should inform states that should legislate. Our present management difficulties, he writes, stem from the paucity of legislation and the 'deaf ears' politicians have turned to scientists.

The two historians who contribute to this volume, Horden and Reuss, and the one political scientist, myself, certainly disagree with such a conclusion. Focusing on the medieval Mediterranean area, Horden argues that the relation to water is always embedded in the political and social structure of a society. For example, the mills around Lucca in the late sixth century cannot be understood without knowing the demography, the power structures, and the property relations existing in that part of Italy at that time.

Examining the development of water that took place in the United States, Reuss roots his understanding of water resources in American political and social values and in American institutions. He distinguishes two pervasive elements in American political behaviour that fundamentally affected water development: the distrust of powerful governments and the belief that power and liberty are fundamental antagonists. This led to an insistence on the dispersion of power among the three branches of government and a reluctance to see decisions made by non-elected bodies.

Turning to the nature of water conflicts, my own chapter challenges the dominant assumptions made by the modern water-war and water-peace schools of thought. Both often rely on indicators of water scarcity as if scarcity and conflict were not both socially constructed.

Two of our contributors have been actors themselves in the field of water management. Ian Byatt was the director of OFWAT, in the UK, endowed with powers he describes in his essay. Stefano Burchi, the senior legal adviser of the development law office at the Food and Agricultural Organization, part of the United Nations, actively contributes to shaping the water law adopted by states as diverse as Uruguay, Mexico, and Tajikistan. Both of these contributors focus their attention on the state and how it should or should not deal with water management. Such an approach flows from the very nature of their occupation, and yet it contrasts starkly with the implications of Horden's conclusions. He deplores the lack of a bottom-up approach in studying water development through history. In my own contribution, I try to understand why researchers tend to adopt a state-centric view in spite of overwhelming evidence of its inadequacies.

The contributions from Sir Ian Byatt and from Stefano Burchi are also significant because they open the way towards building an integrative as well as an interdisciplinary research framework. Integrative research incorporates practitioners as well as scientists in the construction of the research. Understanding how the actors themselves define the issues, which problems they identify, and which solutions they promote is a key step in developing a sound understanding of water governance. These two chapters each give a water practitioner the opportunity to articulate his view of water management.

The political geographer, Maria Kaika, focuses her work on the adoption of the European Water Framework Directive. Far from being state-centric, Kaika explores the role played by supranational bodies, non-governmental organizations (NGOs), and others with parochial interests in shaping water management in Europe.

Finally, Peter Ashton, an environmentalist, takes us back to Africa, from where Edmunds initially drew most of his examples. Ashton explores the issue of sustainable development of water while bringing in considerations such as the spread of HIV/Aids in Southern Africa and the reform of land-ownership practices. His concern for governance institutions that include the transparent engagement of stakeholders echoes that of Maria Kaika about Europe.

This book is therefore a multidisciplinary attempt at exploring water management. It does, however, contribute to the development of an inter-disciplinary approach not least by identifying some of the obstacles in its way. The widely varying views of the authors, even their clashing identifica-tions of what the core issues are in water management, indicate the crucial points that must be addressed if a common interdisciplinary framework is ever to be built.

Of the several themes running through these contributions, development and democracy emerge as key concepts that are perceived very differently within various disciplines.

DEVELOPMENT

Peregrine Horden emphasizes throughout his contribution that power over water is always contested and argues against technological determinism. Technological change, he says, is not a grand secular progress from incom-petence to management in the history of water. He resolutely pleads against a unilinear vision of history that would correlate changes in water tech-nology with phases of history, and perceive each phase as an improvement on its predecessor. Rather, a great variety of strategies have somehow involved water for some use at various points in time. Horden argues that topics such as irrigation and flood control need to be contextualized within the whole spectrum of interactions between humanity and water. For example, he links the relation with sea water to that with fresh water. Sea water proved to be *the* medium of redistribution in Mediterranean history and this affected the uses made of fresh water.

The perception of water development that emerges from Horden's work is clearly not one where more control over water is equated to a greater level of development. The mechanisms that govern the evolution of water use rather lie in the individual strategies developed in the face of the climatic uncertainty of the Mediterranean basin. Increasing crop diversity was the best safeguard against such uncertainty. A variety of agricultures and water uses were developed to that effect, within an overall strategy towards risk embedded in the political and sociological structure of a society.

This historian's perception of water development certainly echoes my own understanding, which is illustrated in my contribution. The same approach to uncertainty that leads individual actors to practise crop diversity also often leads them to trust other forms of social organizations besides the state. This explains the persistence and resilience of communal water management institutions. In circumstances where a state is weak and unreliable, an irrigating farmer will rather trust and obey communal institutions. The latter often tend to be more efficient at rule enforcement and will provide the farmer with more certainty concerning his access to, use of, and allocation of water and the transmission of his rights to access, use, or allocate water. Such institutions have rarely been recognized by the water-war and water-peace literature, which has led to an incomplete understanding of the nature of water conflicts so far.

The unilinear perception of water development as one of growing control over water has been deeply rooted in the foundational literature on water management. Garrett Hardin in his 'The tragedy of the commons' (1968), completely failed to recognize communal property regimes. He identified only public property regimes, private property regimes, and open access, which he wrongly labelled 'the commons'. He portrayed the development of natural resource management as a unilinear progression from open access to either public or private property regimes. In other words, he defined the development of natural resource management as a progression from a free for all, unregulated situation to a situation of greater control over the resource in order to moderate its use.

Elinor Ostrom (1992, 1993) and others responded to Hardin with the common properties movement. Research on customary forms of water management repeatedly dispelled the myth of the unilinear progression from less to more control or less to more technology. It demonstrated that open access cases were extremely rare in the case of water and that communal property regimes regulated most water sources involved in irrigation (Mabry 1996). Disciplinary compartmentalization largely prevented such research from spilling over to the natural sciences, engineering, law, and many practitioners of water management.

While Martin Reuss also roots his understanding of water development in the US in American political and social values and in American institutions, water development appears to be conceptualized very differently in other contributions to this volume. Sir Ian Byatt, for example, offers an account of United Kingdom water management that resembles a Hardin-like evolution from an open access situation to a public property regime and, finally, to a private property regime. Sir Ian Byatt has had a distinguished career, and he understandably relates what Horden would define as a set of complex interactions between society and water to the policies implemented by the Conservative government from 1979 onwards. He details the function of the water regulator once the water infrastructure was privatized in the UK. The regulator was to ensure 'that the companies properly carry out their functions and can finance them'. He emphasizes the fact that the regulator may

consult ministers, but retains final responsibility. Although such a feature
could be described as a democratic deficit, Byatt considers it in a positive
light. In his view it promotes efficient water management through a specific
use of market forces and through increased power in the hands of a non-
elected official who will remain impartial and impervious to popular pressures.

The concept of water management development that emerges from
M. Edmunds's contribution also matches the Hardin-type progression from
open access to some form of property regime that would allow for more control of water extraction. Edmunds does not make privatization a necessity
however, but clearly argues in favour of natural scientists having a greater
role. For him the development of water management should rest increasingly on a 'scientific understanding' of aquifer recharge mechanisms.
Edmunds deplores the fact that an absence or paucity of legislation in many
countries has allowed uncontrolled drilling and abstraction that is endanger-
ing the very renewal of the resource base.

The contributions of both Byatt and Edmunds present a concept of water
management development that is based on efficiency, although their criteria
to define that efficiency vary. Byatt is concerned with economic efficiency:
achieving the highest water service quality at the lowest cost. Edmunds is
concerned with hydrological efficiency: respecting sufficiently the integrity
of the water cycle so that water is not abstracted faster than it is renewed. In
Martin Reuss's discussion of the evolution of American decision-making
concerning water infrastructure, there is a third concept of efficiency. He
shows how the early twentieth century took 'scientific efficiency' to mean a
maximal use of water for distinct human purposes, and it was only after 1936
that American planners emphasized economic efficiency instead. It is inter-
esting to note that the concept of scientific efficiency prevalent among
American planners at the beginning of the twentieth century was quite
different from that which underlies Edmunds' emphasis on respect for the
hydrological cycle. The very definition of scientific efficiency is not an
objective, rational reality that is independent of the natural scientist. Rather,
it is constructed and shaped by the values prevalent in the scientific community at a particular moment in time.

The various concepts of water management development that emerge
from the various contributions to this volume can be broadly classified into
one of two categories: teleological development, and socio-political development. The first considers development to mean an increasingly efficient control by an actor, deemed to be the legitimate repository of power over water
(often the state), exercised over the various aspects of resource management
that are deemed essential by the author. The second category considers that
control over water is always a matter of competition among many actors and
that the outcome changes as the political and sociological structure of

society changes. Water management development in this case is defined as an evolution of the interaction between society and water in a direction allowing equitable access and use to all social groups as well as to nature at a given point in time.

While Edmunds, Byatt, Burchi, and Ashton express a view of water management development that fits the teleological development category, Reuss, Horden, Kaika and I argue in favour of a definition that fits the socio-political development category. This is a direct consequence of the manner in which each of our disciplines defines the issues involved in water management. The teleological perception of water development can allow a reading of history as a progress from a 'worse' situation to a 'better' one, from incompetent management to competent management. The socio-political perception of water development prevents such a reading. Here, the complex relation between water and society fluctuates in relation to an extensive array of variables such as values, power relations, and economic structure that seem at first sight unrelated to water.

Whether, having recognized the various concepts of water development that underlie their theoretical framework, researchers can agree among themselves on a common definition of water development constitutes a crucial test to determine whether they can undertake interdisciplinary and integrative research together. Many interdisciplinary endeavours fail at this stage.

DEMOCRACY

The process whereby water management choices are made and the various perspectives that are included or excluded within this process constitute the core of the debate on democracy and water management. Several radically different views on the articulation of water management and democracy emerge from the various contributions to this volume.

Martin Reuss explores the tensions between expertise and democracy within water management throughout American history. He observes that decision-making dominated by expertise tended to ignore the interests of minorities while decision-making dominated by democratic concerns tended to parochialism. In the United States, tensions between expertise and democratic concerns in the past were often settled by the judicial system. Recently, the development of alternative mechanisms of dispute resolution has allowed an avoidance of the court system.

Reuss distinguishes two pervasive elements in American political behaviour that deeply affected water management: a great distrust of powerful governments and a profoundly entrenched belief that power and liberty are fundamentally antagonistic. The Constitutional Convention rejected

Benjamin Franklin's proposal that Congress have the power to construct canals for fear it would make it too powerful. This led to corporations carrying out water infrastructure development supported by the state governments as if they were government agencies.

There was a long tug of war between the federal government and the states concerning whether water infrastructure works funded by the wider public but having only local benefits were acceptable. Retracing this history, Reuss illuminates the ways in which various local and national actors competed or co-operated, driven by changing needs and values. The need for irrigation water and for hydropower led to the Colorado River Compact in 1922 being concluded among all Colorado basin states except Arizona. Later, the New Deal of F. D. Roosevelt led to the creation of the Tennessee Valley Authority that became a social laboratory. Every time the thorny issue of who was legitimately entitled to make decisions was bitterly debated and compromises that were initially accepted were later challenged. Reuss concludes by reflecting that '[b]y its very nature then, public works tests the resiliency of American institutions and challenges the nation to seek cooperative answers rather than capitulate to a much easier solution: authoritarian direction.'

Similarly, in her contribution, Maria Kaika explores the process that led to the adoption of the Water Framework Directive (WFD) by the European Union (EU) in late 2000. While the EU invited active public involvement in planning river basin management, she raises the question of who was permitted and who was able to participate. The European Commission launched an open call for participation in drafting the directive, but also invited its chosen interlocutors. A group, observes Kaika, needed to know the call existed in order to respond, and those who had 'their man in Brussels' were in a better position to do so. Moreover, few groups could afford to follow the whole process. Big NGOs such as the World Wild Life Fund therefore became very active in the drafting of the Water Framework Directive.

Kaika observes, both in the implementation and in the drafting phase, a strong tendency to replace civil society with NGOs. She points to a democratic deficit, questioning the nature of an NGO and the sources of pressure for social justice. Investigating the disagreement between the European Parliament and the European Council over the content of the WFD, Kaika shows that it embodied the culmination of conflicting social, economic, and political interests involving actors at various scales. She observes, '[w]ater supply projects are no longer one part of a state-led development of the collective means of consumption, they are also opportunities for market development dealt with according to the "laws" of the market economy and regulated through new institutional structures.' The market, she concludes, is defeating state control and we are witnessing the emergence of a new polity with a reconfiguration and rescaling of power centres.

Kaika's reading of the evolution of water management in Europe is one in which the only institution that provides democratic control over water, the state, is losing much of its power. The consequence is that citizens are losing their capacity to exert control over water management. Interestingly, Kaika disagrees with Byatt's suggestion that the newly privatized water industry provided the necessary funds to invest in UK water infrastructure. She argues that between 1989 and 1995 the British government invested £17.9m. to meet the new quality standards set by the EU's directives. Where Byatt sees increased efficiency through privatization and regulation, Kaika finds government subsidies accompanied by a loss of democratic control. Yet Byatt also argues in favour of public participation. Democratic control and participation appear to be contested concepts requiring close definition.

In his contribution, Burchi explores how the recent evolution of national water legislation around the world is providing for water security and water governance, two notions that were in the limelight at the Second World Water Forum and Ministerial Conference held at The Hague in March 2000, the International Freshwater Conference held in Bonn in December 2001, the United Nations World Summit for Sustainable Development in Johannesburg in 2002, and the Third World Water Forum held in Kyoto in 2003. All these were intergovernmental conferences, and the meaning they attributed to the concepts of water security and water governance was clearly articulated through a state-centric prism. Burchi starts by noticing common assumptions among all states concerning water legislation. First, states have been bringing all or most resources under the scope of the government's allocative authority. This has occurred through profound changes in property regimes of various kinds. First, private groundwater rights and riparian rights in surface watercourses have been eliminated in favour of some form of public property in water resources. As a consequence, individuals have only been able to claim and obtain user rights. Secondly, various forms of water trading have been developed by national legislations in recent years. Such a mechanism partially returns to the users the allocative authority vested in the government. Thirdly, societies' relation to water has been monetarized through charging for water abstraction. Fourthly, non-point sources of water pollution have been curbed via the regulation of land use, for example.

Finally, all national water legislations have been bringing in some form of participation by water users, a phenomenon of interest since it clashes to some degree with the trend towards disempowerment of traditional water-right holders. Whose participation are we observing here? In the Spanish Water Act of 2001, notes Burchi, the legislation called for a compulsory formation of water users' groups when aquifers risk over-exploitation. These users' groups have to share groundwater management responsibilities especially in the management and policing of extraction rights. While Kaika showed concern for the participation of citizens in spelling the rules of water governance, Burchi shows that the evolution of national water legislation calls for the participation of water users in the implementation of the law. Such a perception of the nature of participation is probably what allows Byatt to call for participation while approving of the fact that the water regulator is not accountable to an elected body. Participation in water governance is not necessarily synonymous with democratic control of water governance.

Burchi describes the main priority actions agreed by the delegates to the Bonn conference in December 2001, and these provide a good definition of what the conference participants understood to be water security and water governance. The Bonn priority actions insist on equitable access to, and use of water, efficiency of water use, integrity of the ecosystems, effective legal frameworks, and participation of the 'people' and of 'local stakeholders' in managing local water needs and resources. Interestingly, the contributors to this book have dwelt on several of these topics and have shown that they can be defined in a great variety of ways. They have also shown how each of these notions can come into conflict with the others. Who is a legitimate representative of the 'people' and who is a 'stakeholder'? asks Kaika. Efficiency of water use is defined differently in Byatt's and Edmunds's work. The integrity of the ecosystem appears as a dominant concern for Edmunds and Ashton while Burchi explores the legal framework offered by national legislations.

The competing visions of what is water development and what is democratic water management that emerge from this volume offer many challenges for future research in this area. That is one purpose of this collection. They also contribute to a better understanding of why well-intentioned international aid has often fuelled mechanisms of structural violence in the process of attempting to fund water development projects. Structural violence designates the set of mechanisms that, while not causing direct killing, maiming, repression, and desocialization, breeds inequality to such an extent that the resulting misery resembles that caused by direct violence. (Galtung 1990) Practitioners usually apply state-centric standards of management efficiency and definitions of water development that ignore the role of customary knowledge and communal property regimes. These lead to the marginalization of many poor social groups in the decision-making process concerning water management. Similarly, reckless water mining and pollution of aquifers also fuel mechanisms of structural violence that affect the poorest first and foremost. Only an integrative and interdisciplinary understanding of these mechanisms can lead to genuinely improved water management practices that will not benefit some social groups at the expense of others.

REFERENCES

Allenby, B. 'Observations on the philosophic implications of earth systems engineering and management', *Botten Institute Working Paper* (2002).

Galtung, J., 'Cultural violence', *Journal of Peace Research*, 27/3 (1990), 291–305.

Hardin, G. 'The tragedy of the commons', *Science*, NS 162/3859 (13 Dec. 1968), 1243–8.

Mabry, J., *Canals and Communities: Small Scale Irrigation Systems*, Arizona Studies in Human Ecology (Tucson: University of Arizona Press, 1996).

Ostrom, E., *Crafting Institutions for Self-Governing Irrigation Systems* (San Francisco: ICS Press, 1992).

——— 'Coping with asymmetries in the commons: Self-governing irrigation systems can work', *Journal of Economic Perspectives*, 7/4 (Fall 1993), 93–112.

Silent Springs: Groundwater Resources Under Threat

W. M. Edmunds

SPRINGS IN THE VALLEYS

Springs are symbolic of the sustainability of life on earth. Since the earliest times flowing springs have been held as sacred and as a subject of awe and fascination. Subterranean water is identified in the creation myths on Babylonian tablets, where waters above the earth are separated from the 'water of the deep'. The persistence of these creation myths is still reflected in the Arabic word *ain* or *ayun*, which has the double meaning of spring and eye (Issar 1991). Springs were the eyes of the gods. Springs (or fountains) were the focal point of many events in the Bible and other religious texts, and were the subject of veneration, as in Psalm 104: 10, 'He sendeth the springs into the valleys, which run among the hills . . .'

Modern scientific understanding of the origins of spring flow dates from the seventeenth century. The earliest explanations of the hydrological cycle, often termed the *reversed hydrological cycle*, probably stem from biblical sources (Ecclesiastes 1: 7). The unexplained constancy of the ocean volume was accounted for by the return of seawater through the rocks, which then purified them and returned the water to the surface as freshwater rivers and springs. This interpretation of the hydrological cycle persisted through the writings of ancient Greece and Rome as in Seneca's *Quaestiones Naturales* and into the Middle Ages (Tuan 1968) until correctly explained by Edmond Halley (Halley 1691).

In modern society spring waters are valued highly because they still embody an element of mystery and bring us face to face with the subsurface expression of the hydrological cycle or 'groundwater'. There is also traditional belief that spring waters represent a source of perennial pure water. The properties of pure spring water command a high market value and in a world where tap water is (often wrongly) perceived as something less pure, the bottled water image-makers seek after evidence of the purity, longevity, and healing properties of the spring, with a zeal that echoes the reverence accorded to spring waters by early philosophers.

The objective of this chapter is to explore the reasons for the decline of natural springs and the fragility of groundwater resources in general. How much renewable groundwater is there and how do we measure it? How do we identify non-renewable waters and what are the challenges for using these very limited resources? An emphasis is placed upon semi-arid and arid regions where spring flows have traditionally been of great importance and which today face the most acute threats from over-exploitation. Examples are taken from countries in northern Africa—Sudan, Libya, and Nigeria— which illustrate how groundwater resources have evolved over long time-scales, how they have been recharged today and in the past from rainfall and riverflows. Although springs today are a rare commodity in drier regions, they may still feed lakes and oasis regions in parts of the Sahara and Sahel.

FALLING WATER TABLES: CLIMATE CHANGE AND MAN'S INTERVENTION

In contrast to northern Europe and other temperate regions most arid regions were once much wetter places and the onset of the present drier climates, for example in North Africa and the Middle East, started around 4,500 years ago (Gasse 2000). As these changes took place, perennial rivers became intermittent and unreliable and eventually ceased flowing altogether, causing population migrations from inland areas, to the coast and to the major river valleys (Hassan 1997). Springs, however, remained in many areas, nurturing early civilizations and providing reliable water supplies despite periods of short-term climatic change. Some springs were developed into chambered systems and some were excavated to create shallow wells, but human intervention during early human history had little impact on the overall hydrology. The water engineering of the classical era worked with the natural hydrology by canalizing large springs and creating dams and galleries as well as *qanats*, which originated in Persia and spread throughout the Near and Middle East.

Well-drilling technology was known to the ancient Egyptian and early Chinese civilizations and was introduced in Europe by monastic commun-ities in the twelfth century, possibly to offset the drier and warmer climates of medieval times. It was not, however, until the twentieth century that serious disturbance of the subsurface hydrological cycle occurred with the wide-spread practice of borehole drilling; indeed in semi-arid regions this devel-opment phenomenon is restricted to the last half of the twentieth century, when large sedimentary basins were exploited for water at the same time and with similar technology as for their oil reserves.

Stresses on water tables and on traditional spring-fed water supplies have now therefore become acute throughout the arid and semi-arid regions. In parts of China, Mexico, and the USA as well as over much of the Middle East and North Africa, for example, annual falls in the water table of several metres are not uncommon. This clearly spells out that abstraction far exceeds renewal. Many galleries and *qanat* systems, which have flowed for millennia, are now dry. Fragile ecosystems and desert oases, which have supported rich fauna and flora as well as traditional agriculture, are now fast disappearing as water tables are lowered. The finite nature of water resources in sensitive semi-arid regions is also sharply emphasized as human populations increase, leading to a decline in per capita water availability. The current situation therefore reflects a water crisis, created not only by a lack of understanding of the nature of groundwater occurrence but also by the general mismanagement and over-exploitation of water resources over recent decades.

HOW MUCH RECHARGE IS ACTUALLY OCCURRING?

Reliable estimates of the world's renewable water resources remain elusive. Only 2.5 per cent of the world's 1,386 m. km^3 of water are fresh and around 30 per cent of these are stored as groundwater (Shiklomanov 2000). However the vast proportion of groundwater is contained in deep basins, which were replenished during the geological past and renewed only on the scale of 1,000 to 1,000,000 years. Modern rainfall in semi-arid regions can, under favourable conditions, still recharge the system and this topping-up of groundwater reservoirs is balanced naturally by spring discharge including baseflow to rivers or the seas. This renewable groundwater, vital as it is for most semi-arid and arid regions, is generally overlooked in global estimates (Shiklomanov 2000; Gleick 2000). For many regions any renewable water is now significantly overdrawn. In the Gulf region (Bahrain, Kuwait, Oman, Qatar, Saudi Arabia, and UAE) the total replenishable resource to shallow and deeper aquifers is 6.2 billion cubic metres (BcM) against a total usage of 20.6 BcM, causing a large deficit (over 15 BcM), mainly due to large-scale agricultural use (W. Zubairi 2002, personal communication).

Estimation of the amount of groundwater renewal or *recharge* is thus a critical parameter and considered by most hydrogeologists as the most difficult quantity to measure in creating a 'water balance'; much effort has and still is being devoted to this subject, as reviewed recently by Scanlon and Cook (2002). It is generally agreed that, if at all possible, two or more different recharge estimation methods be used. Physical methods have traditionally been used for recharge estimation, but under arid or semi-arid climates these break down, especially due to difficulties in measuring rates of evaporation, which occurs when they are high and rainfall amounts are low and

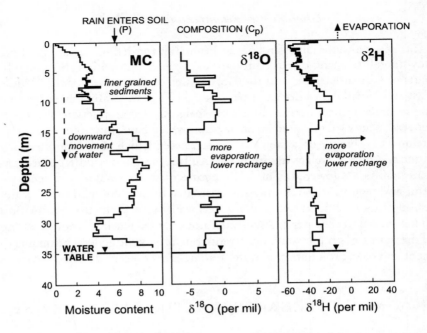

FIG. 1.1 Records of rainfall recharge in the unsaturated zone of the Quaternary sediments in north-west Senegal: this example shows a profile through old sand dunes, which cover a wide area of north-west Senegal. Rain enters the soil and after undergoing evaporation percolates 35m to the underlying water table. The water (moisture content MC) varies between 1 and 9% of the total sand weight, lower moisture corresponding to coarser sand horizons. Tritium (^{3}H) acts as a time marker and the 1963 'bomb peak', derived from thermonuclear weapons testing,

highly seasonal (Allison *et al.* 1994). Surface run-off via streams or wadis is also flashy and unpredictable, being notoriously difficult to measure in remote desert regions. Recharge may take place regionally through the soil (diffuse recharge) or via selected pathways such as wadi channels (selective recharge).

Geochemical tracer methods offer the possibility for improved point and regional estimation in low recharge areas. Such tracers include those chemical elements/species or isotope ratios, which remain unchanged on passing into groundwater, thus providing a record of the input conditions. Geochemical and isotopic tracers serve as 'forensic' tools and the corroborating evidence of more than one technique usually lowers the uncertainty in the recharge estimation (Herczeg and Edmunds 1999). The ideal tracer, or tracers, for the purpose will remain inert during passage into the soil and by downward percolation through the unsaturated zone above the aquifer eventually reaching the water table. Whereas water is evaporated and is *not* conserved, the solutes such as chloride contained in the rainfall are left behind. Thus, the higher the salinity in the residual moisture, the lower the rate of recharge and vice versa, and the use of residual solutes provides a

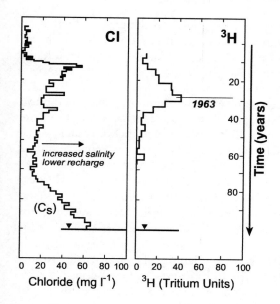

also shows that there is a piston-type displacement (with no bypass flow) of the water. The moisture represents (approx) 100 years of storage.

The water isotopes record the degree of evaporation in the moisture, as does the chloride. The increasing salinity and more positive isotope values show the drier years; these indicators show well the severe drought of the 1970s and 1980s. The rate of recharge (R) may be calculated if the rainfall amount (P), the rainfall composition (Cp), and the composition of sand moisture (Cs) are known (Rd = P.Cp/Cs); in this example the long-term average recharge is 34mm/yr.

sensitive method to estimate the recharge. The chemistry of the rainfall is initially derived from the oceans; these marine solutes 'rain out' further inland but may be augmented by airborne dusts. The rain over continents is therefore very pure, and a prerequisite of the mass balance approach is that reasonable estimates of the rainfall chemistry and amounts are made. Chloride is the most inert of the solutes and is used in preference to other chemical species, which may undergo reactions with soil or rock. The use of chloride for the estimation may then be supported by the use of stable isotopes of water, which become enriched in the heavier isotope as the (lighter) water molecule is evaporated (Clark and Fritz 1997). Tritium, the radioactive isotope of hydrogen, produced during thermonuclear testing also acts as a worldwide marker of the nuclear test peak in 1963. This radioactive isotope decays with a half-life of 12.3 years but may still be detected in the unsaturated zones of some aquifers. The application of these tracers to the estimation of recharge in northern Senegal is illustrated in Fig. 1.1. At this location 34mm of annual recharge are confirmed. It should be noted that this estimate is a robust *long-term* average representing some ten decades of fluctuating rainfall contained in 35m of unsaturated sand. In addition to the recharge estimate, the unsaturated zone also provides a record of the changing recharge conditions and, by inference the changing climate during the twentieth century (Edmunds *et al.*, 2003, in press).

In areas covered by sandy deserts (not an insignificant percentage of the arid and semi-arid regions of the world), point estimates of recharge may therefore be possible from simply measuring salinity on its way to the water table.

However, this is only the start of the process, since for water resources estimation, regional assessments are needed. The geochemical approach may be extended in two ways. First, by drilling multiple profiles at a single location it is possible to offset any errors caused by heterogeneity of the soils or local vegetation. Secondly, by sampling shallow wells across such regions (and there are hundreds of thousands of dug wells across the Sahel for example), it is possible to take samples of water arriving at the water table and thereby obtain a record of the spatial variability of recharge using simply the concentrations of chloride (Gaye and Edmunds 1996; Edmunds *et al.* 2002). In Fig. 1.2, illustrating this approach for the extreme north-east of Nigeria, the circle size represents the salinity (chloride concentrations); those areas with smallest circle size therefore represent the locations and regions that have the highest recharge of the aquifer.

This chloride mass balance approach, especially the use of profiling techniques, has now been used in many arid regions. The evidence so far accumulated (Edmunds and Tyler 2002) indicates that a cut-off point of 200mm rainfall is generally the limit for sustainable *regional* recharge to take place, also that recharge rates in a single terrain may be highly variable. Nevertheless there are exceptions. Higher than average recharge may take place where the surface deposits are composed of coarse sands, but the corollary is that saline profiles indicate no recharge. In addition to the diffuse, regional recharge so far described, preferential recharge may also occur via ephemeral streams or wadis, which may flow for only a few days per year. Wadi systems typically contain coarse sand deposits and allow rapid infiltration, in contrast to percolation through soils where rainwater is overwhelmingly lost by evapotranspiration. Recharge accumulations may therefore take place in proportion to the amount of rainfall, providing small replenishable, if fragile, water resources, which can sustain local populations in climatic regions having mean annual rainfall amounts considerably less than 200mm. Today these small amounts of water will not normally sustain spring flows of fresh water. Any natural discharge from groundwater systems is most likely to take place via inland depressions or via the coast. The small recharge amounts are particularly vulnerable to excessive pumping, although sufficient renewable resources may be available for use if drawn by traditional methods, which take small quantities of water.

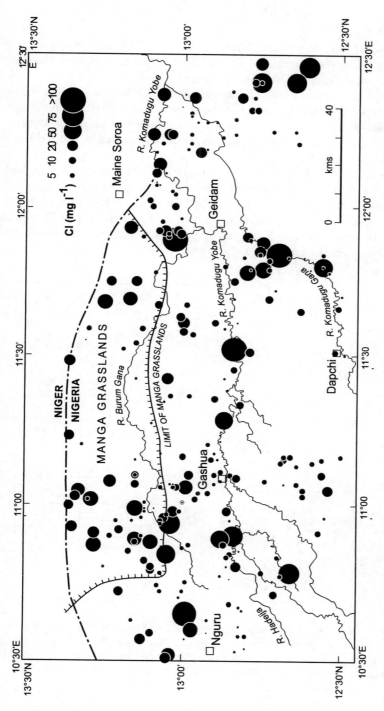

FIG. 1.2 Chloride in groundwater as an indicator of recharge amounts—an example from the north-east of Nigeria (Borno State), where the circles represent the chloride concentrations in individual shallow wells (up to 30m in depth). The circle size is proportional to chloride concentrations and inversely proportional to the amount of recharge.

W. M. Edmunds

WADIS AS LIFELINES

Wadi Hawad is an ephemeral tributary of the River Nile in Sudan, which it joins some 20km downstream of Shendi. This major wadi is joined in turn by a series of smaller wadis (locally termed *khors*), which are fed inter- mittently by monsoonal rains. This area, like most of the Sahel, underwent a severe drought between 1969 and 1990 resulting in a weighted mean average annual rainfall of 154mm against a longer-term average (1938–68) of 225mm. It lies therefore at or below the threshold of replenishable groundwater resources.

Geochemical and isotopic fingerprinting techniques have been applied here to illustrate the relationship between the wadi system and the under- lying groundwaters, as well as the nearby River Nile (Darling *et al.* 1987; Edmunds *et al.* 1991). Some of these studies are focused on Abu Delaig, a typical small town in the Sahel, some 200km north-east of Khartoum (Fig. 1.3) lying on the banks of a small *khor*, that normally flows for short

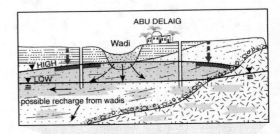

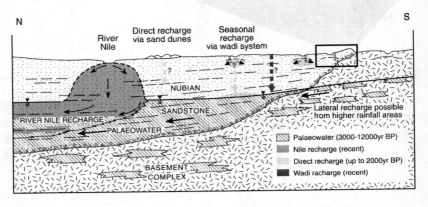

FIG. 1.3 Cross-section of the Butana region of central Sudan between the Nile and the Atbara rivers. Modern recharge in the Nubian Sandstone takes place from the shallow wadis from the intermittent summer rains sustaining the settlements. Apart from this (and recharge from the River Nile) there is no direct recharge and the deeper water is palaeowater and non-renewable.

periods several times per year. Results from this small area show quite well
how the local shallow groundwater has evolved from rainfall and from
surface waters of the wadi, and most importantly, providing a valuable and
sustainable, if fragile, natural resource.

Abu Delaig lies almost at the boundary between the permeable Nubian
Sandstone system (which extends across Sudan and Egypt to the Mediterranean) and the low permeability granitic rocks of the African Basement
Complex, which is encountered at shallow depth (26m) in wells to the south
of the town. The interfluve areas form flat grassland with sandy soil but
often with a clay matrix, which imparts a relatively impermeable surface;
gravel ridges are also common near the basement/Nubian contact. Outside
the town the area is grazed by local or nomadic farmers who rely not only on
the shallow groundwater resource exploited by hand-dug wells (to 26m) but
also on several deep (to 150m) pumped boreholes which were drilled further
north in the mid-to late twentieth century into the Nubian Sandstone.

The cross-section (Fig. 1.3) shows the main elements of the landscape,
with the Abu Delaig khor shown magnified. The water isotopic diagram
(Fig. 1.4) then shows the fingerprints of the different waters participating in
the regional hydrological cycle. This diagram represents a generalization of
many data collected from a field season in 1982–5.

Water contains isotopes of oxygen and hydrogen in different proportions.
The heavier isotopes (oxygen–18, deuterium) become enriched (indicated
by more positive values of $\delta^{18}O$ and δ^2H) by processes of evaporation in
air masses and are also enriched in the equatorial rainfall relative to higher

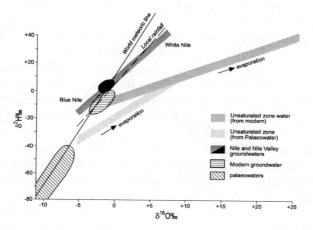

FIG. 1.4 Isotopic evidence is used to provide the fingerprinting of the waters in
Sudan as illustrated in Fig 1.3. Clear distinction is made between the modern
groundwaters which are related to rainfall (or the River Nile) and the isotopically
much lighter palaeowaters.

latitudes. The heavier isotopes also become depleted at higher altitudes as air masses lose moisture passing over continents. Any rainfall falling near sea level and near the oceans should lie close to the World Meteoric Line (WML) which links all global rainfall, the more *negative* values representing higher latitudes (cooler climates), the more *positive* values, enriched in heavier isotopes, the equatorial regions. The slight offset of the local (Khartoum) rainfall line (to the right-hand side of the WML) indicates that moisture forming the rains has undergone some evaporation passing from coast to continent. The River Nile shows a similar trend to the rainfall and its component water sources may be recognized. The White Nile derived from equatorial Africa has the heaviest isotopic signature whilst the Blue Nile derived from higher altitudes in the Ethiopian Highlands is more negative.

This Nile signature indicates that there is local loss or recharge from the River to the north of Khartoum, around Shendi, and that river recharge forms an important contributor to groundwater along parts of the Nile valley. This is an important factor in the water budget of the River Nile elsewhere. It is also apparent from the isotopic composition of these groundwaters that the Blue Nile floods rather than the White Nile baseflow is a more important contributor of groundwater in the water balance.

The groundwaters being drawn from the Abu Delaig wadi and its immediate vicinity lie on or close to the local meteoric water-line. This tells us that there has been little or no evaporation and that the source of recharge must be the wadi floods. These floods (and the heaviest rains) have isotopic compositions, which are at the lighter end of the rainfall line and lie within the groundwater envelope on the diagram. Further evidence of modern recharge is given by the detection of tritium in the groundwaters near the wadi and extending beneath the town.

In contrast, the waters in the unsaturated zone are highly enriched in oxygen–18 but still related to modern rainfall by the intercept on the local meteoric water-line. These waters are also correspondingly saline and indicate that no recharge through the interfluve areas can be occurring under the present-day rainfall regime. The local nomadic population has relied on shallow wells in or adjacent to the wadis for many generations. However, new deep wells were drilled in the area to offset the worst effects of the droughts. Isotopic signatures show very clearly that these deeper waters do not belong to the modern hydrological cycle and must have been recharged under cooler climates. The deeper aquifer therefore receives little or no water from today's rainfall and cannot provide a sustainable resource, only a strategic reserve. In fact it is also apparent from the diagram that where the water table is shallow, losses of the palaeowaters can also occur by evaporation through the unsaturated zone.

Thus the wadi systems serve as lifelines in this and other areas of the Sahel and Sahara as well as in similar regions of the world. In times of drought nomadic populations traditionally knew where shallow water occurred along these valleys, possibly the sites of former springs. As the climate deteriorated over the past 4,500 years, wells were dug instead of springs. In this region examples may be found of ancient wells as much as 100m in depth where well diggers have followed the water table, in decline, over the centuries. Today the wadi systems, which contain permeable sediments, act as conduits for recharge during the few days of heavy rains. This water is still sufficient to sustain a traditional way of life in Abu Delaig and similar small towns and villages. It may still be possible to sustain 'linear' communities via shallow wells along the Wadi Hawad and elsewhere; the deeper wells drawing palaeowaters do not provide an alternative supply, only a valuable, non-renewable, emergency resource.

DESERT SPRINGS AND OASES: SUSTAINED BY FOSSIL WATER

So far we have considered the evidence that small water resources found in springs and wells—even lakes—may still be maintained by finite but small amounts of rainfall entering permeable parts of the earth's surface at the present day. However, the traditional desert oasis areas of the world originate primarily from the running down of the large hydrological systems fed by regional rainfall recharge during much wetter climates. As seen above the last effective wet periods in the Sahara and most of the worlds' arid regions ceased over 4,500 years before the present.

The past hydrology of the continental regions may be inferred indirectly from the geological record. Dated sedimentary and biological remains, preserved in organic-rich layers of lake sediments, contain isotopic and other records, which can help reconstruct the past climate and hydrology (Gasse 2000). For much of the past 100,000 years the climate of North Africa, as well as the rest of the world was cooler than today with extended polar ice-caps leading to sea-level lowering of 100m or more. This was accompanied by a shift in the westerly air flow towards the equator bringing wetter climates to the present-day Sahara. This prolonged cooler and wetter period led to flowing rivers and the formation of lakes and the emergence of springs across the region. Large rivers flowed from the central Saharan mountains of Tibesti and the Hoggar towards the Mediterranean and also to the south of these mountains, evidence of which is visible from air photographs and satellite imagery as the layout of a former substantial drainage system. Recharge to the aquifers took place at a regional scale and the overall higher water tables enhanced the flow of groundwater in the sedimentary basins towards the lowest points of discharge coincident with modern oases.

Evidence exists in the sedimentary record from this time of large ground-water-fed lakes, for example in the oases of Kufra in Libya, and Kharga and Dhakla in Egypt, as well as terminal inland lakes fed by rivers elsewhere.

At the time of the last glacial maximum (LGM) in Europe a period of aridity lasting some 8000 years (22,000–14,000 BP) is recorded across most of the present Sahara and Sahel region. This was then followed by a remarkable period of climatic instability related to the melting of the polar ice, uneven warming and the rapid rise in sea levels around 8000 BP. This timescale was marked by abrupt oscillations of wet and dry periods, related in particular to the variations in the monsoon intensity in that region until the onset of the present-day arid climate. During that time the lake sediments record high rainfall periods with high lake levels lasting on average 2000 years, with shorter dry interludes (Gasse 2000).

These climatic and hydrological events are borne out also by the groundwater record. Groundwaters, like the lake sediments, may be dated using the carbon–14 contained in dissolved bicarbonate (HCO_3) in the groundwater; accurate dating is not easy, however, since the input signal from the radioactive carbon dioxide is difficult to establish for the past centuries and millennia and also the radiocarbon is diluted to different extents by the non-radioactive carbonate minerals contained in the rocks making up the aquifer (Clark and Fritz 1997). Water samples taken from most deep aquifers indicate that they contain no radiocarbon and therefore must be older than 30,000 years. From gases dissolved in these waters it has also been established that during the last glacial period the mean annual air temperatures were some 6–7 °C cooler than today. There is a noticeable absence of groundwaters with dates in the region of 12,000–20,000 years BP, which equates with a cessation of regional recharge to groundwaters during the arid phase coinciding with the LGM.

Following this period of aridity, there is evidence of localized recharge to groundwaters that can be dated to the wet phases between 5000 and 12,000 years BP. Most of these records are of groundwaters, which were replenished by leakage from the larger rivers. In Libya significant recharge from now extinct rivers took place during the period 5500–8500 BP and in parallel with this are excellent records of Neolithic settlements in the river valleys (Edmunds and Wright 1979). These late-stage groundwaters overlie the main freshwater body of late Pleistocene age in a recognizable mound sitting over the older waters; some evidence also is found, where water has been dated at the immediate water table, for small amounts of a regional recharge having taken place during the wet periods. Other recharge from the Holocene is recognized from Niger River near Toumbouctou since this river was then much larger, flooding areas north of its present course (Fontes *et al.* 1991).

THE LAKE DISTRICT OF THE SAHEL

Despite its current aridity as compared with the past, one area of the Sahel, Lake Chad and its region, still contains lakes fed by rivers and springs under the present rainfall regime. Lake Chad is a vast inland basin (2,500,000km^2) with a hydrographic region extending from southern Algeria to the tropical rainforests of Cameroon (Fig. 1.5). It is fed by the Chari and Logone Rivers to the south and the severe drought of the 1970s and 1980s seriously affected the geographical extent of the lake. The lake levels and corresponding surface areas varied from a low of 275.4m and less than 3000km^2 in 1984 to a high of 283.4 and almost 26,000km^2 in 1962 (UNEP 1989). Locally the lake provides recharge to shallow aquifers at times of high stand (Isiorho and Matisoff 1990). This also accounts for the fact that Lake Chad remains fresh; any salinity is lost through the permeable lake bed and the turnover of the lake itself is relatively short, unlike most other inland lakes in tropical regions which become saline.

The Manga area (Fig. 1.5) has been the focus of recent work covering its hydrogeology and palaeohydrology (Carter and Alkali 1996; Holmes *et al.* 1999; Edmunds *et al.* 1999). This 'lake district' is probably unique in the Sahel, although in recent decades only a handful of the lake basins contained any water. As well as the remaining perennial lakes many dry lakes or playas are to be found. The waters are mostly fresh, although brackish or saline lakes also occur. The piezometric surface in the Manga area is about 30m higher than Lake Chad and it is evident that the recharge source is independent from that of the lake. The Yobe River also flows through the area, although its flow has been much reduced in recent years due to upstream impoundment of its headwaters. The mean annual rainfall in the area for the period 1961–90 was 434mm/yr which represents a 30 per cent decrease from the period 1942–60, consistent with decreases seen more widely in the Sahel (Hulme 1992). Oscillations in the lake levels appear to be closely connected with the longer-term, decadal scale climatic change.

The dune fields of the Manga Grasslands were formed during the Late Pleistocene and have been reactivated at times up to and including the present day (Mortimore 1989). The main vegetation cover of the dunes consists of annual grasses with some perennial species. The dune field overlies uppermost sediments of the Chad Formation comprising mainly lacustrine deposits (sands and clays) of Plio-Pleistocene age. There are three major aquifer units of the Chad Formation, the deeper two of which are confined by clays and artesian pressures give rise to overflowing borehole conditions across much of the region (Fig. 1.6). The Lower Zone aquifer has a thickness of 89m at Maiduguri with alternations of sand and clayey sand. The thickness of individual unconsolidated sand units reaches 14m and the grain size varies from fine to coarse. It has been proved at least as far as Lake Chad but

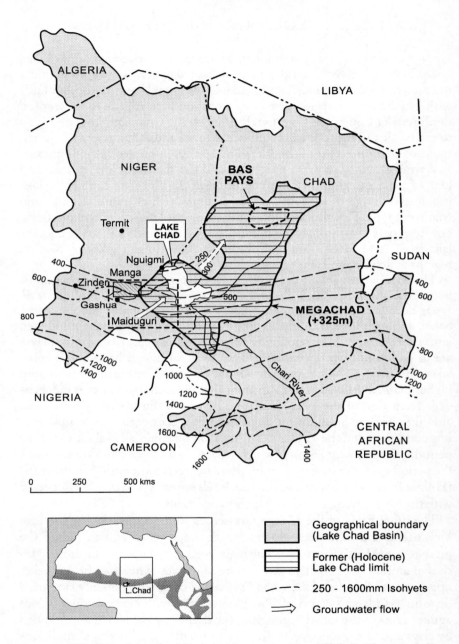

FIG. 1.5 The area of the Chad Basin, with present-day rainfall distribution. Lake
Chad is shown with its former maximum extent during the Holocene (Megachad).
The area of Fig. 1.2 is shown as a dotted outline and the Manga Grassland as a
shaded area.

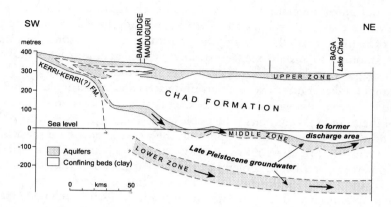

F IG . 1.6 A cross-section through the Quaternary sediments of the Chad Basin. This shows the flow of water beneath Lake Chad towards the former discharge area north-east of the lake. These groundwaters are palaeowaters and are not apparently being recharged significantly at the present day; any modern sustainable supplies are to be found in the upper aquifer.

has only been exploited in a few boreholes on account of its depth. The Middle Zone aquifer reaches a maximum thickness of 31m and forms the main water supply unit. The sands are fine to very coarse, poorly graded, and mostly uncemented. The initial hydraulic gradients established during the extensive studies of the early 1960s showed a general flow direction towards Lake Chad. The Upper Zone consists of interbedded sands and clays and forms an aquifer of variable quality, which underlies the whole region. The main questions arising which affect the water resources are: (1) is recharge taking place to sustain the lakes (and the shallow groundwaters) at the present day? (2) what is the age of groundwater at depth in the artesian Middle and Lower Zone aquifers of the Chad Formation and is the water renewable? and (3) is there interconnection between these two aquifer systems and also with Lake Chad?

Estimates of modern recharge have once more been made using geochemical (chloride profile) methods (Edmunds *et al.* 2002). Moisture samples obtained from unsaturated-zone profiles gave recharge rates through the dunes ranging from 16 to 30mm/yr. The spatial variability of the recharge to the Manga Grasslands was calculated using chloride mass balance results from 360 regional shallow wells over an area of 18,000 km^2 (shown in Fig. 1.2). The regional Cl balance indicates a regional recharge of 43mm/yr, suggesting either that additional preferential flow is taking place over and above that from the unsaturated zone, or that the regional recharge represents inputs from earlier, wetter periods. These recharge estimates are encouraging and compare favourably with those from hydraulic modelling in the same area, suggesting that the regional recharge rates are much higher than values previously published. These values indicate that around 10 per cent of rainfall is rapidly entering the permeable sandy sediments in sufficient amounts to sustain seepage springs, which maintain lakes and the shallow aquifer, at least during periods with higher rainfall.

Samples taken from the deeper groundwaters and in particular from the Middle Zone aquifer were analysed for chemical, isotopic parameters as well as for their noble gas contents. The noble gases (Ne, Ar, Kr, Xe) dissolved in groundwater remain chemically inert but their ratios depend on the atmospheric temperature; these ratios therefore provide an indicator of whether the climate was warmer or cooler than the present day at the time the rainfall entered the soil. A comprehensive understanding of the evolution of the aquifer system, extending from its outcrop and probable recharge area to the south near Maiduguri towards and beneath Lake Chad, is therefore provided.

Groundwaters in the aquifer are fresh and apparently extend beneath the lake to a depth of 300m or more in both the Lower and the Middle Zone aquifers (Fig. 1.5). However, the radiocarbon dating shows that the groundwaters were recharged over a very narrow time interval from 24,000 to 18,600 years BP. Moreover the noble gases dissolved in the groundwater, which accurately measure the temperature at the time of recharge, show that the mean annual air temperatures at the time the water moved into the aquifer were some 7 °C cooler than today. Tracing waters back to the area of outcrop indicates that there is no modern water entering the deeper aquifer.

The present-day water resources therefore have been influenced strongly by the past hydrology of the region. During much of the late Pleistocene surface drainage took place via the Bahr el Ghazal valley to a lower area (the 'Pays Bas') to the north-east of the present-day Lake Chad, in the Chad Republic. During this time the regional groundwater flow also took place beneath modern Lake Chad. Following a period of extreme aridity, corresponding to the glacial maximum in Europe, the Holocene in the Sahel region was marked by extreme monsoon activity during which a huge lake (Lake Megachad) formed which extended across much of the region (Fig. 1.5). The high lake levels (some 43m higher than the present day) covered the discharge zone of the aquifer and prevented through-flow.

The artesian groundwaters in the Chad Basin are therefore not being replenished to any significant extent at the present day and the artesian area of the aquifer is rapidly decreasing. Today the unconfined upper aquifer (Fig. 1.6) forms the main storage for any modern rainfall recharge, although as described earlier the rates of replenishment may vary with climate change. In addition the upper aquifer is discontinuous, due to the nature of the alluvial sediments and well yields vary from place to place. For the future, reliance will need to be placed increasingly on the shallow groundwaters of the upper aquifer, as practised by previous generations. Thus, after fifty years of using the bonus of the overflowing borehole waters, the population will need to consider how to make best use of the shallow aquifer where this exists, since in some places near to Lake Chad the clay cover inhibits recharge. In this regard, the spring-fed lakes area of the Manga Grasslands and the nearby plains of the ephemeral Yobe River are shown to have relatively high rainfall recharge and contain sustainable resources of groundwater.

QUANTITY AND QUALITY: A DOUBLE THREAT

The aridity of the past 4,500 years has taken its toll on the quality as well as the quantity of available and renewable groundwater. Although fresh groundwater springs still feed some of the desert oases most of these are now silent due to the lowering of water tables in the past few decades near to discharge zones. The discharge areas and associated wetlands have become dessicated and native vegetation, especially commercially valuable palm groves, are sustained only by irrigation and the area of native vegetation near to oasis margins has been reduced.

This scenario illustrates the fragile nature of the desert ecosystems and their underlying water supplies. Much of the area in the vicinity of oases had already become naturally saline as water tables declined over the past four thousand years and evaporation from shallow water tables produced brackish or saline waters. Human intervention, especially the practice of irrigation with deeper freshwater may have produced transient improvements in quality but the large accumulations of salts are easily remobilized to affect the shallow groundwater quality. Nevertheless the quality of groundwater discharging as springs or exploited from depth may be quite variable. Beneath Kufra in southern Libya water at a depth of 700m in the Nubian Sandstone aquifer has a salinity (expressed as total dissolved solids or TDS) of 162mg/l, which is three times lower than the mineral content of London tap water. This water feeds (or rather fed) via springs a series of hypersaline lakes containing brines with a salinity of 119,000 mg/l TDS (total dissolved solids), as well as producing solid rock salt (halite) in the vicinity of these lakes. Such freshwater occurrences support fragile ecosystems, which are important, not least for the migrating bird populations. A dilemma exists therefore in relation to the exploitation in these areas. Should the non-renewable fossil water be pumped for short-term gain at the expense of the oasis wetlands and ecosystems and with the risks of salinity problems? It is often argued that it is economic to pump groundwaters from these areas—beneath the salt lakes or *sebkhats*—since by lowering the water tables that water which would otherwise be lost by evaporation can be intercepted for use. This debate needs to take into account the fate of the wetlands so that in any future management scheme the water-table drawdown may be compensated. A healthy ecosystem is usually an indicator of a healthy water resource.

Not all deep groundwaters are as fresh as those at Kufra. The Chotts region of southern Tunisia and central Algeria is an area of salt-pans and salt lakes, fed by discharge from vast aquifer systems owing their origins mostly to rainfall that fell up to 100,000 years ago in the Atlas Mountains. This groundwater emerges naturally as springs but in recent decades deep drilling of boreholes to depths up to 2,000m has tapped larger volumes of this water resource. From such depths the water emerges at high pressure with artesian heads up to 40m, temperatures up to 70 °C and with salinity up to 2,000mg/l. This water is at best marginal for drinking but is used widely for agriculture, especially date farming. The higher salinity waters originate from mixed lithologies (impure sandstones with some carbonate minerals) compared with the very pure Nubian Sandstone and have acquired dissolved sulphate and chloride during transit. These discharges over time have created large salt lake areas. Moreover as the thermal water emerges it cools down causing precipitation of gypsum and calcite. This produces severe scaling of the pipeworks and a strong potential for salinization of the soils and quality deterioration in the less saline shallow aquifers. Thus, despite the potentially large reserves of fresh or slightly brackish water, quality considerations may create the greatest constraints to development.

Salinity problems encountered near to the oases have to a large extent been avoided by large-scale development schemes, which have tapped into the fresh groundwaters through well fields in remote areas of the desert away from oasis centres, either to develop new settlements or, as in Libya, to export supplies to the coast. The Great Man Made River (GMMR) in Libya produces groundwater from two huge well fields, which is then piped over 500km to supply the coastal cities, including Tripoli and Benghazi, as well as boosting the area of irrigated land. These waters are of good quality, but as discussed earlier, these are palaeowaters which must be regarded as non-renewable and can serve as only a temporary answer to water supply shortages. Careful management of these reserves, as with the oil, is required to provide for future generations. Technological solutions requiring large energy inputs are unsustainable as far as water is concerned; humans need water to survive but not necessarily oil.

Apart from the general distribution of salinity, additional quality problems may sometimes be found in natural groundwaters that are not part of the present-day active water cycle and which have been in prolonged contact with the rocks. As water moves away from its recharge source its atmospherically derived oxygen is progressively reduced by reaction with either organic matter or aquifer minerals. Thus, the aerobic groundwaters will with time become anaerobic and allow iron to become soluble. Most of the large reserves of fresh waters in pure sandstones beneath deserts remain aerobic and traces of oxygen remain. Nevertheless, build-up of other elements such as chromium, arsenic, and uranium may occur in certain lithologies and careful checks for similar metals (which form oxy-anions) need to be made. A further problem is caused by the stability of nitrate under the aerobic conditions and this is widely reported from desert groundwaters (Edmunds and Gaye 1997; Hartsough *et al.* 2001). This nitrate most probably originates from the leguminous vegetation that formerly covered desert areas at times of climatic optimum.

From the above it is clear that the natural quality of groundwater related to the aquifer geology and hydrological setting can provide unique management problems for desert groundwaters, even before human pollution is considered. Thus, the scarcity of fresh groundwater in many regions of Africa and the Middle East, as well as other semi-arid regions, demands their careful protection. Human intervention provides many ways of reducing the availability of fresh groundwater without actively contaminating it. The major issues surrounding water pollution are beyond the scope of this paper. Nevertheless, the pressures arising from agricultural practices, rapid urbanization, tourism, and ever-increasing volumes of waste-water and contaminated land, produce locally severe contamination problems for the underlying aquifers especially where these are unconfined and most vulnerable.

CONCLUSIONS

Groundwater, emerging as springs, touches deep roots in human consciousness. References to life-giving properties of fresh water and springs may be found in the writings of ancient civilizations, the great religions, poetry, and music. The loss of spring waters and the loss in quality—in fact their silencing—still stirs passions. Spring flows and their associated fresh groundwater may also be regarded as a sensitive barometer of healthy ecosystems. Forty years after Rachel Carson wrote her classic book on the state of the environment (Carson 1963), it would be wise to regard our natural water status as both a cultural and environmental indicator.

In the above examples it has been shown how little of any modern rainfall may actually reach the water table and that many water bodies in semi-arid and arid regions are fed by 'fossil' water. The small renewable amounts may still be adequate to sustain traditional rural communities. However the uncontrolled drilling of new boreholes with the installation of electrical pumps in recent decades in these areas has more often than not led to rapid water table decline and overdraught of water.

An improved understanding of the nature of groundwater is required and here a greater dialogue between scientists and society as a whole is required. Improved estimates of the amounts of renewable groundwater resources are still needed in all countries, but especially in the more arid regions. It is important to be able to recognize whether or not we are actually 'mining' groundwater and, if so, to develop water management strategies that match the scientific evidence. We need to ensure that we are able to manage the renewable water resources to cater for the water needs of future generations. However, if it is proved that the resources are non-renewable, their abstraction needs to be carefully managed under the realization that the settlement in such regions is non-sustainable.

Traditional water uses need to be re-evaluated to see how these well-tried methods may best be adapted, aided where necessary by technological improvements to deal with scarce and erratic rainfall and limited recharge; water harvesting, well-design efficiency, and not least conservation need to be applied in these areas. However for many countries the short-to medium-term solutions lie in a reform of the practice of using scarce groundwater wastefully in agriculture. Water quality issues are equally important and care must be taken not to write off the scarce amounts of renewable water (or the non-renewable groundwater) by ignorance or carelessness. Although the application of science together with traditional wisdom can raise awareness of the problems, water scarcity for many people, as pointed out by others in this volume, is a political issue and reforms in society are necessary to ensure that more springs do not fall silent and that groundwater in general may continue to sustain communities.

REFERENCES

Allison, G. B., Gee, G. W., and Tyler, S. W., 'Vadose-zone techniques for estimating groundwater recharge in arid and semi-arid regions', *Soil Science Society of America Journal*, 58 (1994), 6–14.

Carson, R., *Silent Spring* (London: Hamish Hamilton, 1963).

Carter, R. C, Alkali, A. G., 'Shallow groundwater in the north-east arid zone of Nigeria', *Quarterly Journal of Engineering Geology*, 29 (1996), 341–55.

Clark, I. D., and Fritz, P., *Environmental Isotopes in Hydrogeology* (Boca Raton: Lewis, 1997).

Darling, W. G. *et al.*, 'Sources of recharge to the Basal Nubian Sandstone Aquifer, Butana Region, Sudan', *Isotope Techniques in Water Resources Development* (Vienna: IAEA, 1987), 205–24.

Edmunds, W. M. and Gaye, C. B., 'High nitrate baseline concentrations in ground-waters from the Sahel', *Journal of Environmental Quality*, 26 (1997), 1231–9.

—— and Tyler, S. W., 'Unsaturated zones as archives of past climates: towards a new proxy for continental regions', *Hydrogeology Journal*, 10 (2002), 216–28.

—— and Wright, E. P., 'Groundwater recharge and palaeoclimate in the Sirte and Kufra basins, Libya', *Journal of Hydrology*, 40 (1979), 215–41.

—— Fellman, E., and Goni, I. B., 'Lakes, groundwater and palaeohydrology in the Sahel of NE Nigeria: evidence from hydrogeochemistry', *Journal of the Geological Society London*, 156 (1999), 345–55.

—— *et al.* 'Sources of groundwater recharge at Abu Delaig, Sudan', *Journal of Hydrology*, 131 (1991), 1–24.

—— (2003 in press), 'Groundwater as an archive of climatic and environmental change. The PEP III traverse', in R. W. Battarbee, F. Gasse, and C. E. Stickley (eds.) *Past Climate Variability through Europe and Africa*. (Dordrecht: Kluwer).

—— 'Spatial and temporal distribution of recharge in northern Nigeria',' *Hydrogeology Journal*, 10 (2002), 205–15.

Fontes, J.-Ch. *et al.*, 'Palaeorecharge by the Niger River (Mali) deduced from groundwater geochemistry', *Water Resources Research*, 27 (1991), 199–214.

Gasse, F., 'Hydrological changes in the African tropics since the Last Glacial Maximum', *Quaternary Science Reviews*, 19 (2000), 189–211.

Gaye, C. B., and Edmunds, W. M., 'Intercomparison between physical, geochemical and isotopic methods for estimating groundwater recharge in northwestern Senegal', *Environmental Geology*, 27 (1996), 246–51.

Gleick, P. H., 'The changing water paradigm. A look at twenty-first century water resources development', *Water International*, 25 (2000), 127–38.

Halley, E., 'An account of the circulation of the watry vapours of the sea, and the causes of springs', *Philosophical Transactions of the Royal Society*, 192 (1691).

Hartsough, P. *et al.*, 'A 14.6 kyr record of nitrogen flux from desert soil profiles as inferred from vadose zone pore waters', *Geophysical Research Letters*, 28 (2001), 2955–58.

Hassan, F. A., 'Nile floods and political disorder in Early Egypt.', *Third Millennium BC Climate Change and Old World Collapse*. NATO ASI Series I/49 (1997), 1–24.

Herczeg, A. L., and Edmunds, W. M., 'Inorganic ions as tracers', in P. G. Cook and A. L. Herczeg (eds.), *Environmental Tracers in Subsurface Hydrology* (Boston: Kluwer, 1999), 31–77.

Holmes, J. A. *et al.*, 'Holocene landscape evolution of the Manga Grasslands, Northern Nigeria: evidence from palaeolimnology and dune chronology', *Journal of the Geological Society London*, 156 (1999), 357–68.

Isiorho, S. A., and Matisoff, K., 'Groundwater recharge from Lake Chad', *Limnology Oceanology*, 35 (1990), 931–8.

Issar, A. S., *Water Shall Flow from the Rock: Hydrogeology and Climate in the Lands of the Bible* (Berlin: Springer-Verlag, 1990).

Mortimore, M., *Adapting to Drought: Farmers, Famines and Desertification in West Africa* (Cambridge: Cambridge University Press, 1989).

Scanlon, B. R., and Cook, P., 'Theme issue on groundwater recharge' *Hydrogeology Journal*, 10 (2002), 3–4.

Shiklomanov, I. A., 'Appraisal and assessment of world water resources', *Water International*, 25 (2000), 11–32.

Tuan, Y. F., *The Hydrologic Cycle and the Wisdom of God* (Toronto: University of Toronto, Department of Geography, 1968).

United Nations Environmental Programme (UNEP), *The Lake Chad Conventional Basin. A Diagnostic Study of Environmental Degradation* (New York: United Nations, 1989).

Zubairi, W., personal communication, 2002.

2

Water in Mediterranean History

Peregrine Horden

I AM a historian, not a hydrologist, and it is well known that historians tell stories. So let me begin with one that opens up some of the themes I would like to consider. The date is around AD 400. The scene is the fringe of the small provincial town of Nola in Campania, Italy. Here was to be found the shrine of St Felix. It had recently been constructed in the middle of an elaborate complex by the bishop of Nola, Paulinus. Fountains gurgled in the courtyards of the basilica, offering natural refreshment. They also symbolized both the church's and the saint's therapeutic powers—the water of life, the rivers of paradise, baptismal regeneration. Rain-collecting cisterns initially met the requirements of the shrine, but these proved inadequate. The shrine became in effect parasitic on the water supply of Nola.

This supply arrived both through the grand *Aqua Augusta*, overall some 96 kilometres long, and from a small aqueduct that started in the hills of Abella. Now Abella was a very small nearby town that took its water into a reservoir through pipes from high mountain ridges and released the surplus down an aqueduct that both supplied Nola and irrigated the surrounding estates. The aqueduct was refurbished by the Abellans so that it fed St Felix's shrine first and Nola second. The Nolans, however, felt deprived by the suburban complex to the point that they rioted. Yet another aqueduct, a disused one, had to be restored by the Abellans to appease the inhabitants of Nola, even though Bishop Paulinus had already been trying to persuade the Nolans that, by dividing their water with the saint, they reaped unexpected rewards, not just of a spiritual kind. The surrounding fields were better watered than they had been before; the area was better fed. Paulinus wrote in the poem (*Carmen* 21) that is our principal evidence for the local conflict: 'Where rough stones lay arid in bare fields, there is now the pleasant transformation of greenery on the watered turf . . . Felix . . . has also brought here to your city the fountains that flow from heaven' (Walsh 1975: 195–201, at p. 200; Trout 1999: 192–4; Squatriti 1998: 13–14).

Much is implicated in that story: the convoluted politics of water management; the complexity of urban water supplies; the interdependence of local pipe-work and major aqueducts; the links between agricultural irrigation and the supply of civic drinking water. There was nothing new about this in the early Middle Ages. The aqueducts of imperial Rome, to take the most obvious example, fed not only baths and dry throats; by fraud or design they helped the water-intensive horticulture on the city's periphery (Hodge 2002; Purcell 1996). Civic provision is not separable from irrigation— although aqueducts were perhaps essential only for the megalopolises of antiquity. Both Rome and Nola illustrate the impossibility of separating display from practicalities (Paulinus was trying to echo the fountain that astonished visitors in the atrium of old St Peter's in Rome), of separating luxuries from essentials, even of defining luxuries without anachronism. The Nolan vignette shows, above all, the need for a cultural history of water rather than a narrowly technological one—a history with a place for religious and other values. The values may not be universal or even readily accepted; they may have to be asserted, as St Felix did through his spokesman the bishop. Yet we cannot miss them out in our understandable preference for a secularist account of water and its management. Water, as another early medieval bishop, St Isidore, would later put it, can cure or sicken, make plants grow, wash dirt away, quench thirst, and remove sins (*Etymologies*, 13. 3; Squatriti 1998: 8–9). Yet Isidore knew about the water cycle, or at least about the twin water cycles—one atmospheric, the other subterranean—by which seawater was purified and returned, against gravity, to the hills. There was no separation in his world-view between the cultural, the scientific, and the technological (to use our own terminology). If we are looking at local, sophisticated systems of water delivery we should think of the water supply of medieval baptisteries (used perhaps only at Easter and Pentecost) (Magnusson and Squatriti 2000) as well as at modest estates such as those around Nola. In similar spirit, when surveying modern types of water use we should embrace those sacred landscapes of our own time that are greediest for water—golf courses.

I come back to my opening vignette. Felix was a dead saint who needed a living bishop to defend his interests. Living saints, too, were represented as managers of water. I have elsewhere suggested that some saints reported as slaying dragons were managing the threat of malarial waters in a conceptual way. They and the communities they protected were, between them, focusing the threat into an image that, though terrifying, was at least specific, and could thus be tackled (Horden 1992). There are, though, many more— literally—down-to-earth examples. The territory of Lucca provides another Italian vignette and introduces a different kind of hydraulic landscape. The saint is Fredianus. The time is now the late sixth century. Floodwater from the River Serchio and its tributaries was a perennial hazard. Artificial channels 'embroidered' the plain (the wonderful verb of Paolo Squatriti

(1995), to whom modern scholars owe their introduction to this hydraulic saint). Lined with hedges and trees they demarcated property. Some were for drainage, others for irrigation. They supplemented the many natural streams that fed the Serchio. Water mills also punctuated the landscape. Their ownership was shared—some by groups of aristocrats, others by associations of smallholders, others by the collectivities of churches and monastic houses. The mitigation of flooding was generally achieved through the cooperation of cultivators and citizens. And that serves to introduce the theme of who controls what, who owns what, in water technology. The Luccan locals' biggest project in this direction was to prepare a new course for part of the Serchio. They could not, however, make the diversion work. The saint was said, at least in hagiography, to have intervened. He prayed by the riverbank, and then, with a small hoe, traced a new course for the Serchio in the ground. The waters leapt into their new course. As at Nola, some people were disadvantaged by this strategy. Fredianus was very unpopular in the nearby village of Lunata. Power over water is always contested, especially when unevenly applied, as it almost always must be. Recall the etymology of rivalry—*rivales*, those who share a watercourse. Fredianus, the great *rivalis*, experienced rivalry: he had to leave the area. Perhaps the waters of Lunata had dried up because of his hydraulic thaumaturgy.

These examples are Christian. Classical parallels involve sages. The pre-Socratic philosopher Empedocles altered the course of two rivers to purify a third, the noxious exhalations from which were poisoning the city of Selinus (Diogenes Laertius, *Lives of the Eminent Philosophers*, 8. 70). Here is water management of yet a further kind, in the service of public health (Horden 2000). Empedocles' success was so complete that the citizens acclaimed him a god. He leapt into the flames of Mt Etna to demonstrate his divinity: alas, he had been misled. Empedocles was hardly the only ancient figure to be remembered as a hydraulic expert. The drainage of (actually transient) lakes, pools, or swamps has been attributed to heroes, sages, or gods: Hercules for the Stymphalian plain in antiquity, Plato (more surprisingly) for the Plain of Konya in medieval Byzantium (Horden and Purcell 2000: 412).[1] It is as if a great figure is needed to resolve difficulties with water when local initiatives have failed. There has been a perpetual dialogue between the dramatic interventions and the quieter, more sustained, local struggles—a dialogue that I want to stress because one side, that of great interventions, is usually given too much prominence. Models for (or reflections of) a succession of rulers' great drainage schemes run, chronologically, from the artificial lake that watered vineyards, orchards, and vegetable gardens created on the eastern shore of Lake Van by an eighth-century BC king of Urartu, to Mussolini's battle for the plains.

---

[1] Fuller documentation of much that follows may be found in Chapters 6–10 of this book.

Water itself, not just the virtuosi among those who managed it, has of course been the focus of cult—sacred wells, springs, lakes, or swamps. These have their fearful obverse in the various watery abodes of demons. Such abodes include the Roman baths, again showing the impossibility of dividing our subject up into country and town, nature and culture. The great anthropologists were aware of the importance of water cult (Horden and Purcell 2000: 420–1). Frazer gave the facile explanation: water is sacred because it is scarce. Robertson Smith was more sophisticated. He observed that water is less venerated in the desert than in lands where, because it is more abundant, it makes agriculture possible. The religion of water, here as in my earlier Christian vignettes, is at least in part a religion of irrigation, whether natural or man-made.

I have begun in this way, with religion, rather than with the details of aqueducts, drainage systems, or water-lifting devices, or with enquiries into the legal history of water rights, to unsettle our perceptions of what might be involved in the history of water and the environment. In this chapter I have two messages to impart. One is that the obvious topics such as irrigation and flood control need to be seen within the whole spectrum of interactions between humanity and water; they must be treated, in a sense, ecologically. The other message is that, in the project of understanding that ecology, social relations and cultural values—relations of power, symbols of life— need at least as much attention as technological developments.

Let me dwell on the latter message for a moment. To begin to support that assertion about the primacy of culture and value, we may return to an aspect of my second opening vignette. We cannot understand the mills around Lucca without knowing the demography, the power structures, and the property relations of the area. The technology by itself tells us little. Technological determinism of even the most limited kind is no use. It fails, as I hope to show, because it ignores the social context. It also fails because its proponents keep getting their facts wrong. For Marx of course, mills produce feudalism (he referred to hand mills but the general point is unaffected). For Marc Bloch, in a classic article of 1935, still the starting point of modern research, the mills of the Carolingian age eventually produce a new type of independent artisan to run them. They begin to demarcate ancient from medieval society (Bloch 1935). Now the mills of the Luccan plain in the sixth century refute Bloch (and Marx) both in their chronology and in the variety of ways in which they were owned—strictly speaking, part-owned. Moreover, ancient evidence accumulates that the usual picture of very limited milling in antiquity—because of slavery or some other textbook cliché—is probably misleading. Even the argument that such rude mechanical things as mills were unlikely to be noticed in the documents of high culture and civic values needs revision. In AD 324–6 Orcistus in Anatolia (in the upper Sangarius valley near Amorium) was seeking recognition as a city.

It adduced both its baths and its 'abundance of water mills on the down flows of the streams that flow past the town' (Chastagnol 1981). As a civic amenity, the watermill could be set alongside the bath. How very un-Roman we might have thought—quite wrongly. A contrast between antiquity and the Middle Ages, on which we thought we could rely, simply disappears. In the *diffusion* of technology, which is historically more significant than the *invention* of technology, values are trumps. And those values require careful elucidation.

I have one final example to offer in this tirade on behalf of cultural history. Anthropologists have documented the selective resistance of some southern European villagers to economically advantageous change (Horden and Purcell 2000: 291–2). The Greek village of Vasilika, in the Boeotian plain beneath Mount Parnassus, offers the best example. When Ernestine Friedl conducted fieldwork there in 1955–6 the inhabitants numbered 216. The desire for new technology could be conditioned by customary attitudes to the city, rather than by so-called rational economic considerations. Women in Vasilika wanted their houses to have electricity because all houses in Athens had it. They also wanted running water. Yet those without the latter continued to draw their water from the well by hauling the buckets up manually. A windlass would obviously have made their task far easier, but, because the windlass had nothing to do with the technology of the city, it was not seen as desirable.

All my material so far has been circum-Mediterranean. My brief was to consider the early interactions between humanity and water as a background for the studies of present and future to follow. I have narrowed that brief for obvious reasons of space and competence. I have also done so because the longer view is conducive to a teleology that I hope to avoid: *Vorsprung durch Technik*, as the car advertisement has it. Imagine how the longer view might be set down. A little prehistory would supply tentative beginnings. Drainage channels for bananas and taro cultivation in Highland New Guinea date from *c*.4000 BC. Or recall those prehistoric rock engravings from Brescia that may evidence irrigation in the Alpine valleys by the end of second millennium BC (Delano Smith 1979: 178). Perhaps hunter-gatherers often diverted water to wild harvests. Next, the scene must shift to 'the dawn of history': to the Near East and 'Wittfogel country', the lands of 'oriental despotism', hydraulic bureaucracy, and the origins of civilization, across Mesopotamia, the Indus valley and Shang China; then to archaic Greece, Etruscan cuniculi, Rome's aqueducts, and so forth. The temptation is to correlate changes in water technology with phases of history and to see each phase as an improvement on its predecessor. Yet again, though, the facts get in the way and frustrate the implicit teleology. For example the noria—the chain of pots driven by a donkey walking in circles to raise water from a well—was for long comfortably associated with Islamic sophistication in

moving water, but is now attested in pre-Islamic Syria, at Apamea on the
Orontes (Balty 1987: 9–10).

Instead of the breathless progressivism to which the longer and wider
view tempts us, I want to concentrate on the Mediterranean, especially
Mediterranean Europe, though with some reference to Islamic Spain. This
enables me to draw on my collaboration with Nicholas Purcell in *The
Corrupting Sea*, and also on the splendid new study of Grove and Rackham,
The Nature of Mediterranean Europe (2001), as well as on the writings of
Paolo Squatriti, already mentioned, the author of pioneering studies of the
aquatic world of early medieval Italy.

Concentration on the Mediterranean, additionally, permits me to return
to the first message I briefly conveyed earlier and to support it in more detail.
The message was that obvious topics such as irrigation and flood control
need to be seen within the whole spectrum of interactions between humanity and water. I want to mention at least some of those interactions, general-
izing so far as one can across antiquity and the Middle Ages into the end of
pre-modern times in the region, around the early nineteenth century. I
should stress, by way of preface to the account, that I am projecting these
interactions against a *relatively* unchanging backdrop. That means, first, that
I am disregarding climate. Climatic warm periods and ice ages should ideally
come into a fuller presentation. It is, however, hard enough to detect any
particular effects of global warming in the modern Mediterranean. To quote
Grove and Rackham (2001: 126): 'since [1850—before which records are too
few and untrustworthy] *there is no evidence for a general wetting or drying
trend*', though temperatures have risen since 1980. The effects of earlier
changes on the resilient, flexible microecologies I am about to consider will
have been even harder to pick up. How can we calibrate the impact of
long-term, secular change in a world of constant fluctuation and frequent
extremes? A single deluge can do more damage than centuries of slightly
increased precipitation.

Climate has not been the only focus that specialists have selected for a
grand narrative of Mediterranean ecological history with major implications
for the story of water. Two of them I shall merely mention and then set on
one side. It is enough to refer to Vita-Finzi's brilliant, although now unsustainable, hypothesis of widespread more or less simultaneous alluviation of
the Mediterranean valleys after the Roman Empire—the 'Younger Fill'
(Horden and Purcell 2001: ch. VIII. 3). A yet more embracing and more
pessimistic account of long-term ecological change in the Mediterranean
has been developed since the seventeenth century, and a version of it is still
to be found in J. R. McNeill's 1992 study of Mediterranean mountains
(Grove and Rackham 2001: ch. 1). This is a narrative of the environmental
degradation—even desertification—that has been taking place in the
Mediterranean since supposedly lush classical times, at least on European

shores. Forests promote rain, according to the underlying argument (they do not, in fact). The inferred sequence in the 'drying out' is this: deforestation, erosion, maquis, steppe, bare rock, depopulation. Grove and Rackham have blown every link in this chain apart. They show that each stage had multiple causes, rather than just the preceding one in the list, and was not nearly as serious as the pessimists argue—until, perhaps, we reach the twentieth century. For Grove and Rackham, one of the greatest agents of erosion in Mediterranean Europe is the modern bulldozer. 'The major changes between the aboriginal landscape and that of the nineteenth century [the beginning of modernity with the advent of steamships and railways] had, where sufficiently known, already taken place by the Iron Age' (p. 80).

This is my relatively unchanging backdrop. Change within those parameters of Grove and Rackham—disastrous change such as that brought by floods or drought—is normal. It is not part of a grand secular degradation. And in much the same way, I want to suggest, technological change is not part of some grand secular progress from incompetence to management in the history of water.

I turn now to Mediterranean diversity. It is facile to say that a region exhibits diversity: the equivalent of saying that a historical period was an 'age of change'. None the less, the Mediterranean is arguably more diverse than other comparable regions, more diverse than its Middle Eastern, sub-Saharan, and, even, European neighbours. The solid geology is more varied locally and the regolith more unstable; the plant life is diverse to an analogous degree; the landscape exhibits more dramatic contrasts. To walk across Crete is, as Grove and Rackham say, to travel in effect from Wales to Morocco (p. 11). This diversity is equally characteristic of climate. Of the supposedly typical Mediterranean climate—hot dry summers and mild rainy winters—twenty-four different subdivisions are recognized in Southern Europe alone. Eight of them can be found within Albacete province, not the most mountainous part of Spain; that is, within an area no bigger than Crete. Extreme variations in rainfall, both geographically and from year to year, are also commonplace. Add to this the variations in anthropogene effect and in culturally derived perceptions of landscape. The Mediterranean is an immense collection of microfundia, or microecologies—always changing and interacting, defined not by the enumeration of their physical features but by the totality of relationships between humanity and nature. Even the technology can be more diverse in this area than elsewhere. Medieval Spanish horizontal mills are apparently more various than those of Britain or Germany: they have a variety of extra features that deliver water to the drive wheel under pressure.

Microecologies are almost all polycultures: in the Mediterranean past, few have needed to be reminded of the adage about not putting all one's eggs in a single basket. Risk-spreading, flexibility, adaptation, resilience—these are characteristic. Of all the variables water is the most important for food production—on which I now want to concentrate, having alerted you to such other consumers of water as baptisteries and shrines.

Consider the variety of water needs in Mediterranean production. The supposedly Mediterranean triad of cereals, grapes, and olives will not do as a summary. It ignores legumes for a start. Then the olive may be absent because of cold winters or cultural predisposition. Irreducible to any pithy formulation, the Mediterranean is an area in which the seemingly extraordinary may approximate the norm. Grove and Rackham offer a modern, although (for our purposes) still illustrative, list of the ingenious recourses of southern European producers (p. 12): sugar-boiling in Motril, acorn-eating in Estremadura, pig-drying in Alpujarra, esparto-twisting in south-east Spain, madder-growing in Provence, boar hunting and chestnut milling in the Apennines, oat-growing on the Macedonian serpentine, cotton-picking in Boeotia and Crete, quail-gathering in the Mani, and so on—and no mention in their list of rice or weed cultivation, or of pastoralism. Imagine the variety of water needs represented in even a short excerpt from such a list. These forms of production are not, however, to be thought of as dominating localities. No genuine specializing is possible; rather, each bizarre practice is an extra element in a portfolio always being reviewed.

The historical Mediterranean landscape should not be regarded as divisible into prime and marginal: almost all land has its marginal aspect; everywhere is risky. Fertility is not a given, it has to be created. It is, virtually, a social construct. A successful microecology must be characterized by overproduction for storage (more than a good year's needs)—storage against bad years, or against commandeering by the powerful as rents, tributes, plunder. To talk of subsistence strategies is for that reason misleading. No one aims at mere subsistence—at least, no one who wants to survive a bad year. Nor can anyone realistically pursue the ideal of autarky: self-sufficiency is an ideal, but not a practicable one. Production for exchange may always be necessary to make up local shortfalls. In the Mediterranean, we might even say, every crop is potentially a cash crop.

Now add water to the already teeming picture. Microecologies use water in numerous ways that are far removed from the big human interventions of irrigation and drainage. Many microecologies have the sea as an extension of their hinterlands. Water is central to Mediterranean history in that sense too: the sea has been *the* medium of redistribution. Seawater has had other advantages, somewhat neglected by historians. For example: the uses of seawater for adding piquancy to the taste of wine or as preventive medicine against seasickness (recommended by medical writers). These constitute a byway of the subject that can only be alluded to here. There have, though, been other ways in which seawater history impinges on that of fresh water, and these must be given some attention.

First, let us look at wetlands (Horden and Purcell 2000: ch. VI. 5). Ecological historians have wholly underestimated wetlands. So many of them have been reclaimed during the last two centuries that they are not seen as a canonical part of the Mediterranean landscape. Nevertheless, they should be. The coast of Albania—as, until recently, protected from modernity by communism—reveals something of the earlier Mediterranean coastal norm. Wherever a seasonal watercourse backs up behind beach deposits, wherever the accidents of topography render a valley floor slow to drain, or wherever fault-lines have created intermontane basins (zones of inland drainage), there is a potential wetland. Wetlands range in degree of saturation from perennial pool or lake to marsh that dries out in summer. The flood plains of any perennial river can provide an extended chain of environments constantly shifting according to the river geomorphology, the length of winter inundation, and the permeability of the soil, and creating what ecologists call 'hydraulic disturbance patches'. Mutability is characteristic: each season brings change, at a different rate according to the weather, the layout of reed-bed and lagoon, salt-flat and sand-bank; and with these seasonal changes come changes in vegetation and wildlife.

The productive uses of wetlands are multiple. The vine can thrive in very moist conditions. Artisan production of textiles can flourish in wetland environments, as at medieval Tinnis, a city covering an island in the seaward part of the Nile Delta. In eighth-century Palestine Saint Willibald was impressed to see sheep immersed in a wetland at the source of the Jordan to cool them during the dry heat of summer. Each facet of the wetland's usefulness can be managed in a more or less intense and complex fashion, and no one technique is characteristic of any particular historical period. Sophisticated wetland management was for example important in the economy of the Roman coastal villa. Here aviaries, fish-ponds, and game-reserves formed a stylized and controlled extension of the gathering practices of a normal wetland environment.

Next, let me mention fishing, equally underestimated in the historiography of Mediterranean economies (Horden and Purcell 2000: ch. VI. 6). In the sea there are microecologies, which are geographically and chronologically varied. Fish crops vary, but in times of glut can be sold or in times of stress consumed locally. In the years 1600–5 tunny exports from Palermo ranged between 675 and 1,353 tonnes. Another vignette brings the religious theme back into the picture. The first statue on the right as a visitor entered the sacred precinct of Apollo at Delphi was that of a great bull dedicated in about 480 BC by the people of Corcyra (modern Corfù). As the inscription on the base declared, the city dedicated the statue to Apollo out of a tithe of the takings from the sale of a miraculous draught of tunny. Fish shoals on this scale were an unexpected and erratic gain, an eagerly exploited treasure-trove. As community wealth, the tunnies may be compared with the Athenians' contemporary windfall of a newly discovered vein of silver. The catch, its sale, and the record were all systematic, the result of organized collective action. Most importantly, the fish were an asset capable of being turned into disposable wealth—and this already at the beginning of the fifth century BC. The fish clearly fetched prices out of proportion to their nutritional value: there was a lively market for them. Such yields were sometimes public property, a further indication of their potential significance. From the later first century AD, the elder Pliny gives us a vivid illustration of a publicly managed Roman fishery, the *stagnum Latera*, in the territory of Nîmes. There, he alleges, tame dolphins, for all the world like sheep-dogs, helped to catch the shoaling mullet.

I now come to dry farming in this survey of Mediterranean water uses (Horden and Purcell 2000: ch. VI. 8). That will lead us into what we really want, but which I have deliberately delayed: irrigation. Dry farming used to be simple, at least for Mediterranean scholars who never actually had to do it: moisture was retained by biennial fallow with frequent tilling to reduce evaporation and weed transpiration. A strikingly central role in Mediterranean nutritional systems has been played by cereal crops, and by wheat and barley in particular, satisfying perhaps up to some 65–70 per cent of nutritional needs. This centrality has encouraged the assumption that there was, in the pre-modern Mediterranean, a relatively uniform system of dry farming. Of course, a significant overall dietary contribution does not entail a standardized system of production. It is the revolution of recent decades in ethnoarchaeology—the combination of archaeology with ethnographic enquiry—that helps us to see these supposedly standard practices as only part of an enormously more varied and flexible response on the part of producers. The foodstuffs that form the core of a nutritional system may themselves be more diverse than appears at first sight, for instance because they require rather different labour regimes. Cereals are, by comparison with other Mediterranean food plants, relatively low in their demand for labour. Yet they are still susceptible to a great many quite different production relations. Moreover, given Mediterranean climatic and topographical diversity, it would be madness, if not suicide, to attempt to fulfil so important a dietary need with a single way of growing a single type of crop. In different corners, you grow different plants. It is even possible to grow them in the same corner. That is, one of the principal ways of spreading the risk is mixed cropping: sowing mixed seed, or the more systematic planting side by side of different species (intercropping), or using the same land subsequently for a second or third crop during the season. And cereals are far from being the whole story. Unless the powerful enforce a dangerous, because fragile and risk-heavy, monoculture, we have to add to the emergent tableau pastoralism, legumes, olives, chestnuts, acorns, figs . . . In other words, we have to return to the sort of list of unusual resources I quoted earlier. Even in the reasonably predictable conditions of Nile-watered Egypt, records from around 116–15 BC show only 55 per cent of the inner core of irrigable land around the village of Kerkeosiris under wheat; 11 per cent was devoted to lentils, the same again to beans, and 10 per cent to vetch.

In a wide-ranging account of the original domestication of cereals, Andrew Sherratt (1980) has argued for 'an often protracted phase of small-scale surface- and groundwater based horticulture which ultimately differentiated into hydraulic and rainfed systems of both extensive and intensive kinds'. The setting of Mediterranean production indeed more closely resembles the garden than anything that we should normally associate with the word 'farming'. As the geographer Vidal de la Blache earlier put it (1926: 129): 'here the garden rather than the field was the focus of sedentary life'. The usefulness of Mediterranean orchard/garden agriculture as part of the portfolio even of mainly nomadic pastoralists can be seen in the apparently most unpromising conditions of the volcanic mountains of the southern Sinai desert. Here it was introduced by the community of the monastery of St Catherine's. Until fairly recently the Jebaliyah Bedouin, using flood and well irrigation, grew a huge variety of fruit trees in hundreds of small plots, and maintained vegetable gardens in about half of them; cereals were also grown around the fringes.

Irrigation is now at last coming into the picture of the *coltura promiscua* that has been Mediterranean production. We see it, and the way in which irrigated water may be distributed in time-shares, in another passage in the elder Pliny describing an oasis in Roman Tunisia, modern Gabés. A fantastic picture, undoubtedly; yet it does capture the polycropping that is characteristic of Mediterranean garden production.

A spring [Pliny writes] provides abundant water for a space of some three miles in each direction; it is generous, but still assigned to the inhabitants by fixed allotments of time each day. Beneath a great palm-tree here, there grows an olive, beneath that one a fig, under that a pomegranate, then a vine; below the vine wheat is sown, with legumes in between and here and there leaf-vegetables, all in the same season, all reared beneath the shadow of another cultivated plant. . . . The most amazing feature is having two vintages each year from double-cropping vines. (*Natural History*, 18. 188–9)

The locus classicus of Mediterranean irrigation is of course the Valencian *huerta*. Even such scholars as careful with their generalizations as Grove and Rackham see the Spaniards as an irrigation-minded people. Irrigation is not, though, a reflex action, a national characteristic: it is a choice made in specific circumstances at specific times. In Spain, certainly, the traces of the distinction between watered garden and dry farming may be very old. From the copper age in south-eastern Spain, the carbon–13 content of charred seeds has suggested to some archaeologists that beans were irrigated while cereals were not (Grove and Rackham 2001: 77 n. 23). In the Middle Ages it was the cereals that would be irrigated in the *huerta*. So such categories change over time.

How, then, do we write the history of those changes? Let me stay with Spain at least long enough to use it as the springboard for some incautious generalizations of my own. I do not want to go into the techniques of drainage systems such as those of medieval Valencia, described in English so well by T. F. Glick (most recently Glick and Kirchner 2000). The mistake as I see it is again one of technological determinism. We tend to treat presumed innovations in technique as the key to major differences between one irrigation regime and another, assigning primacy to the 'hardware' rather than the people. And we tend to see these changes as coincident with major cultural changes. So the whole of social history becomes determined by its technological base. Against those such as Glick who stress the independence of irrigation in Al-Andalus from its classical precursors, I would suggest that the evidence is not clear enough for us to make that separation. There may not have been continuity: there was, perhaps, more comparability than the Islamicists argue.

From technology, back to people and their techniques. The first general point I want to make is this. Irrigation and drainage should be seen in the same light as all the other strategies that I touched on earlier. They should be seen, that is, as part of a large and changeable repertoire of adjustment to risk and environmental opportunity. They should be seen as subject to the same degrees of periodic 'intensification' and 'abatement' (as they have been called) that is characteristic of all Mediterranean production.

The second point is that, if irrigation and drainage are seen in these terms, within the context of the whole spectrum of possibilities, they are cut down a little in size and importance. Dry production—of innumerable different kinds—has been the obvious choice in so many parts of the Mediterranean for so much of the region's history. Landscapes dotted with watermills or threaded with filtration galleries can be richly documented from some periods and places. Yet there has been, as I have tried to show, so much else going on, sometimes so poorly evidenced as to fall almost beneath the historian's notice.

The third point is that, when there actually has been irrigation, it has overwhelmingly been for the amelioration or regulation of winter, not for what has been called the invention of summer (i.e. as a growing season). It has nearly always obeyed Mediterranean seasonality. Most parts of the region do not have the perennial water supplies needed, and needed in vast quantities. Spring-fed or oasis-based summer production is highly localized and unusual. Anything more requires a perennial river. Yet consider the obvious example of a perennial river: the irrigation does not turn out to have functioned quite as we might have expected. The management of Nile water through gravity-fed irrigation networks has been practised since the very distant past (the date is still controversial). The contrast between the arid climate and the abundance of the Nile was extreme by Mediterranean standards and gave the Egyptians pre-eminence among ancient peoples for their skill in irrigation. Even in Egypt, however, the cultivation of orchards and gardens throughout the year was a practice only gradually extended across the landscape. The main contribution of the Nile was not to provide water for extra production in months when most Mediterranean regions experienced the dry season: without the Nile, Egypt could not support agriculture at all.

My final point concerns social relations. Wittfogel (1957), with his hydraulic bureaucracies supporting oriental despotisms, casts a long shadow. I am not sure that his model is now accepted as applying anywhere—the Fertile Crescent, the Indus Valley, China. For example, what has often been taken as a Wittfogelian bureaucracy in eleventh-century Buyid Iraq is probably nothing of the kind. The thousands of employees of the supposed 'ministry of water' (*diwan al-ma*), with which the Iraqi state has been credited, are found in a very strange mathematical treatise, almost untranslatable in places, and almost certainly not a government document (T. F. Glick, personal communication). Mediterranean irrigation systems are still less amenable to analysis in 'despotic' terms. They are better seen from the bottom up, as it were; better seen from the local initiatives, which the powerful may co-ordinate and aggregate, rather than, Wittfogel-fashion, from the top down. That is to an extent true even of ancient Egypt. The state was more interested in making sure the labour was deployed at the right time than in managing the irrigation network across the landscape. It is also true of our best-analysed medieval example. From medieval Islamic Spain, the remarkable fieldwork of Miguel Barceló and others has identified a set of transformations across the Iberian landscape, a set of managerial interventions that made it possible for many microregions to become interlocked. It is not unreasonable to adopt Barceló's terms 'hydraulic landscape' or 'hydraulic space' for this kind of effect, which has also been identified as a 'macrosystem' of irrigation (Horden and Purcell 2000: 251ff.). Yet Barceló insists on interpreting that landscape from the social and economic perspective. He associates it with a centralized fiscal system, but one that is dependent on maintaining the autonomy of many (relatively) small producers and of eliminating competition from 'señores de renta'. Whether the social basis for smallholder irrigation in Islamic Spain was tribal, as Barceló urges, and whether that form of organization can straightforwardly be contrasted with a feudal one, is a question I would rather avoid here. 'Tribal' and 'feudal' are adjectives that historians and ethnographers are increasingly chary of using. I would simply suggest that, in reaction against Wittfogel, we should not go to the other extreme of finding cooperative peasant utopias in Andalusi irrigation networks. The primary producer-irrigators who maintained their independence were the lucky ones.

I began with power relations in early medieval Italy and have come to something similar in medieval Spain. So, like the donkey going round in circles to drive the *sakiya*'s chain of water pots, I feel I should stop. I do not know what lessons, if any, this exploration of the pre-modern Mediterranean past may have for hydrologists looking to the future in the region. Between Wittfogel's saying in effect 'size matters' and the Mediterranean small producer's response, 'small is beautiful', I have—obviously enough— favoured the small producer. My further message is that a religious, cultural, social, and ecological study of water use in all its variability is more helpful than a primarily technological account. That, I am sure, is a message that hydrologists do not need to hear because they know it already.

REFERENCES

Balty, J.-Ch., 'Problèmes de l'eau à Apamée de Syrie', in P. Louis, F. Métral, and J. Métral (eds.), *L'Homme et l'eau en Méditerranée et au Proche Orient IV: L'Eau dans l'agriculture* (Lyons: GS Maison de l'Orient, 1987), 9–23.

Bloch, M., 'Avènement et conquêtes du moulin à eau', *Annales* (1935), 538–63.

Chastagnol, A., 'L'Inscription constantinienne d'Orcistus', *Mélanges de l'École Française de Rome (Antiquité)*, 93 (1981), 381–416.

Delano Smith, C., *Western Mediterranean Europe* (London: Academic Press, 1979).

Glick, T. F., and Kirchner, H., 'Hydraulic systems and technologies of Islamic Spain: History and archaeology', in Squatriti (2000), 267–329.

Grove, A. T., and Rackham, O., *The Nature of Mediterranean Europe: An Ecological History* (New Haven: Yale University Press, 2001).

Hodge, A. T., *Roman Aqueducts and Water Supply* (London: Duckworth, 2002).

Horden, P., 'Disease, dragons and saints: The management of epidemics in the Dark Ages', in T. Ranger and P. Slack (eds.), *Epidemics and Ideas: Essays on the Historical Perception of Pestilence* (Cambridge: Cambridge University Press, 1992), 45–76.

—— 'Ritual and public health in the early medieval city', in S. Sheard and S. Power (eds.), *Body and City: Histories of Urban Public Health* (Aldershot: Ashgate, 2000), 17–40.

—— and Purcell, N., *The Corrupting Sea: A Study of Mediterranean History* (Oxford: Blackwell, 2000).

Magnusson, R., and Squatriti, P., 'The technologies of water in medieval Italy', in Squatriti (2000), 217–65.

Purcell, N., 'Rome and the management of water: environment, culture and power', in J. Salmon and G. Shipley (eds.), *Human Landscapes in Classical Antiquity: Environment and Culture* (London: Routledge, 1996), 180–212.

Sherratt, A. G., 'Water, soil and seasonality in early cereal cultivation', *World Archaeology*, 11/3 (1980), 313–30.

Squatriti, P., 'Water, nature and culture in early medieval Lucca', *Early Medieval Europe*, 4 (1995), 21–40.

—— *Water and Society in Early Medieval Italy, AD 400–1000* (Cambridge: Cambridge University Press, 1998).

—— (ed.), *Working with Water in Medieval Europe: Technology and Resource Use* (Leiden: Brill, 2000).

Trout, D. E., *Paulinus of Nola: Life, Letters, and Poems* (Berkeley: University of California Press, 1999).

Vidal de la Blache, P., *Principles of Human Geography* (London: Constable, 1926).

Walsh, P. G. (trans.), *The Poems of St. Paulinus of Nola* (New York: Newman Press, 1975).

Wittfogel, K., *Oriental Despotism* (New Haven: Yale University Press, 1957).

3

The Development of American Water Resources: Planners, Politicians, and Constitutional Interpretation

Martin Reuss

To understand the development of American water resources, one must first look at American political and social values and American governmental institutions. Even a cursory examination shows the lasting influence of decisions and attitudes moulded as the country took its first hesitant steps as a republic. Historian Joyce Appleby (2000: 249) has argued that the first generation of Americans bequeathed 'open opportunity, an unfettered spirit of inquiry, [and] personal liberty' to future generations—qualities, we might note, that often introduce an element of uncertainty into public administration. But if we extend the analysis a bit, we might not only gain an appreciation of the many challenges facing water resource developers, but also illuminate a fundamental question facing democratic nations: to what extent should human liberty be constrained in order to provide and manage a human necessity—water.

Beyond Appleby's observations, one notes at least two pervasive elements woven into American political behaviour. The first, the inescapable, element is distrust of powerful governments. Power corrupts, the first Americans agreed without much hesitation, and the challenge was how to minimize that corruption, how to ensure that good men will not be enticed to do evil, and how to disperse power to minimize oppression. Loudly over the years, Americans continue to proclaim their distrust of big government; even popular presidents generate scepticism when they appear to reach for increased power and authority. Only as a last resort, and then with resignation, not enthusiasm, as during the Great Depression, do Americans turn to the national government to solve their problems (Kelley 1989: 30–1; Wills 1999). The result can be truly impressive: Grand Coulee and Bonneville dams, locks and dams on the Upper Mississippi, the California Central Valley

Project, and the Los Angeles flood control system all came out of depression-era politics, but, as I will argue, all are aberrations in the story of American water resources.

The second element, almost as pervasive as the first, is that power and liberty are fundamental antagonists. The dispersion of power among the three branches of government purposely sets power at war with itself rather than with 'life, liberty, and the pursuit of happiness'. Each branch would be allowed only sufficient power to discharge official duties, and a system of checks and balances would guard against abuse (Wood 1969: 150–61). Recoiling from British monarchism, the constitutional drafters took special care to try to prevent executive branch intrusions into the duties of the other two branches. This was a system that, regardless of its merits, made implementation of rational planning enormously difficult, as water developers soon appreciated.

Political attitudes were one thing; government structure was another. And here the Founding Fathers developed a system that guaranteed further complications. They fashioned a republican form of governments within the government. A century later, young political scientist Woodrow Wilson (1887: 221) thought that this structure posed the principal challenge to American administration. Few water resource planners would disagree. Republican government, it must be remembered, began in the states, not in the new national capital; delegates to the Continental Congress delayed business so they could go home and participate in state constitutional conventions. The formation of these state governments may have excited Americans more than the latter formation of the union itself (Wood 1969: 128), and the American Constitution explicitly guaranteed to each state a republican form of government (Article IV, Section 4). Once the United States achieved its independence, many Americans pondered how citizens could owe allegiance to two governments, two legislatures, simultaneously. Were the states and national government partners or were the states meekly to accept national supremacy? No one at the Constitutional Convention quite knew what to expect from this layer-cake of powers (or was it a marble-cake, twentieth-century political scientists later debated), and numerous, contrary explanations emerged of what the delegates had actually achieved (Elazer 1969; Scheiber 1966). In no area did the confusion become more manifest or disruptive than in internal improvements, especially in water projects that crossed state lines.

The term 'internal improvements' came to mean many things to the citizens of the young republic. It included roads, canals, schools, lighthouses, fortifications, and even technological innovations—almost anything that seemed to provide security and promote the economy. Gradually it came to mean something a bit more specific, though still covering (forgive the pun) a large amount of ground: it applied to what we now call 'infrastructure', and water transportation was a central concern. Benjamin Franklin had proposed at the Constitutional Convention that Congress have the power to construct canals, but opponents won the day, fearing that Congress would become too powerful (Albjerg 1932: 168–9). In fact, the term 'internal improvements' cannot be found in the American Constitution, an obstacle for those seeking affirmative authority for federal involvement in public works. But neither did the Constitution proscribe the activity, which meant to internal improvement advocates that the function lay legitimately within federal authority. This ambiguity not only produced a constitutional quagmire for internal improvements, but it provided a platform upon which larger issues of the role of government and the nature of liberty could be debated. In short, the internal improvements issue amplified and sharpened the debates about the very nature of American republicanism. By any other name, it continues to serve that function to the present day.

Given Americans' distrust of government and emphasis on personal liberty, America's first politicians, and all the generations following, confronted the difficulty of promoting economic growth without expanding governmental authority. One answer was the corporation, a device that actually predated the Constitution but in the age of internal improvements became much favoured. As presumed promoters of the public good, they effectively became agencies of government (Maier 1993: 55). In this way, legislatures could support economic and political development without necessarily involving tax revenue. The fact that individual incorporators might thereby profit aroused little concern. The more important point was that corporations brought together sufficient capital to launch an enterprise, whether a canal or a municipal water system. Even if a number of these ventures brought forth charges of corruption, internal improvement advocates ceaselessly trumpeted the moral and intellectual gifts stemming from public works, as though canals were spiritual as well as economic enterprises. To complaints that corporations disenfranchised people and led to the inequitable distribution of wealth, champions argued—somewhat quaintly in light of what subsequently emerged—that corporations were nothing more than little republics that were eminently suited for the United States (ibid. 58–69). For better or worse, the victory of the corporation in American life was almost as revolutionary as the victory of republicanism itself, and the alliance between government and corporations became a hallmark of American economic development. Government was not to replace business, but was to support and, within certain limits, protect it.

George Washington and other Federalists had ardently hoped that corporations might provide the capital and means to build internal improvements to bind the nation together and transcend local interests, perhaps leaving overall planning to the national government. But the chance slipped through their hands. The structure of Congress assured that state interests in internal improvements would prevail over national interest. There would be no national board, no national planning. Rather, Congress would periodically pass Rivers and Harbors Acts that generally reflected parochial politics. To stimulate states and the private sector, Congress also provided a percentage of funds obtained from the sale of public lands in new states to finance roads and canals (the three and five per cent funds dating back to 1802) and voted to turn over certain lands to states for reclamation (Swampland Acts of 1849 and 1850) (Goodrich 1960; Harrison 1961: 67–88; Hibbard 1965: 228–898; Larson 2001: 5–7). In a few cases, too, Congress might vote to subscribe to canal stock or even grant land to a company—a practice that presaged the enormous land grants given to railroad companies as they extended their lines across the continent later in the century.

In many cases, river and harbour bills met overwhelming resistance from the executive branch, mainly because of concerns over constitutionality, yet one more issue that reverberates through much of America's history. The specific points usually centred on presumed lack of constitutional authority to construct works of mainly local and even private benefit or works on rivers that were not clearly navigable. Pre-Civil War Presidents either practised an inconsistent policy towards public works or became adamant opponents. Thomas Jefferson objected to federal involvement because it would empower the government at the expense of the common man, burden the taxpayer, lead to projects benefiting one location at the cost of another, and enrich men at public expense. James Madison, who had been an early internal improvements advocate, on the evening before his departure from the White House vetoed the Bonus Bill that would have provided funds for public works, declaring that the Constitution did not empower Congress to appropriate money for public works without an 'inadmissible latitude of construction' (quoted in Larson 2001: 68; see also Albjerg 1932: 170). James Monroe at first thought that Congress could appropriate funds for public works but agreed with Madison that the federal government had no authority to construct the projects. Later, he determined that Congress might construct public works after all but only for those projects that were 'national not state, general not local,' a clarification that left the proverbial barn door open for defining local and general (Albjerg 1932: 171; Larson 2001: 183).

Andrew Jackson saw himself as a friend of internal improvements, but he feared the extension of federal power, sought a clarifying constitutional amendment on the appropriate national role in internal improvements, and admitted, like Monroe, that, while the federal government could appropriate money for truly national projects, it could not actually construct the projects itself. Strict constitutional constructionist James K. Polk vetoed every rivers and harbours bill sent to him. He even went to his office on the last day of his administration with a veto message in hand should Congress try to pass an internal improvements bill at the last moment. Abraham Lincoln, a young Whig congressman from Illinois, succinctly captured the problem in his denunciation of Polk's veto of the 1848 Rivers and Harbours Act. 'The just conclusion from all this is that, if the nation refuses to make improvements of the more general kind because their benefits might be somewhat local, a state may, for the same reason, refuse to make an improvement of a local kind because its benefits may be somewhat general. A state may well say to the nation, "If you will do nothing for me I will do nothing for you"' (quoted in Kemper 1949: 112). In the following decade, Franklin Pierce vetoed five rivers and harbours bills on the grounds of unconstitutionality. As Civil War erupted in the land, the political and philosophical jousting over water projects remained as short of resolution as ever (Kelley 1989: 31; Larson 2001: 183; Albjerg 1932: 171, 176).

Given the fears, hopes, and questions facing the early American republic, it is little wonder that it saw no successful implementation of co-ordinated public works administration. Perhaps more surprising is that this became a permanent condition in the United States. Funding issues, sectional friction, and constitutional questions invariably posed insurmountable barriers. Prior to the Civil War, the federal government attempted twice to develop and implement a national programme of public works. The first was the well-known Gallatin Plan. At the request of Congress, Secretary of Treasury Albert Gallatin proposed in 1808 an ambitious network of roads and canals connecting the Eastern seaboard with the interior and a coastal water route to shorten distances between major Atlantic seaboard cities. Gallatin argued that the federal government should construct internal improvements that provide 'annual additional income to the nation' but are beyond the capacity of private entrepreneurs to build (Gallatin 1808: 5). His formulation harkens back to Adam Smith's *The Wealth of Nations* and anticipates the term 'National Economic Development' that appeared in twentieth-century economic jargon. However, his effort fell victim to lack of funds (both private and public), New England opposition to the Jefferson Administration, and, finally, growing preoccupation with real and apparent British threats to the United States, which eventually resulted in war. It is also of more than passing interest that Gallatin himself agreed with President Jefferson that his plan could never be efficiently realized without a constitutional amendment (Pross 1938: 10).

The next great attempt occurred in 1824. President's Monroe's vacillation, the growing clout of new states interested in waterborne commerce, and a favourable Supreme Court ruling (*Gibbons* v. *Ogden*) that sanctioned federal control over interstate commerce, including rivers, based on the Commerce Clause of the Constitution, allowed passage of the General Survey Act at the end of April 1824, after weeks of acrimonious debate. The act carried largely because of support from the Middle Atlantic states (except

Delaware) and the new states west of the Appalachians. It authorized the President to use the army engineers to survey (not build) roads and canals (not rivers) that may be deemed 'of national importance in a commercial or military point of view, or necessary to the transportation of the public mail'.

Once the bill passed, Secretary of War John C. Calhoun organized a Board of Engineers for Internal Improvements to determine which routes should be surveyed among the scores suggested. Like Gallatin's plan, this programme could have become the beginning of a great nationwide, co-ordinated system of internal improvements. Instead, once projects were surveyed, they became subject to the same parochialism in Congress that had doomed earlier, similar ideas, and Congressmen continued to introduce pet projects for funding despite a contrary recommendation from the army engineers or the absence of a survey altogether. Its planning role severely diminished, the Board of Engineers languished, and a reorganization of the Corps of Engineers in 1830 provided an excuse for its abolishment. Six years later, Congress repealed the General Survey Act, partially a response to the legislature's own abuse of the act, using army engineers to survey potential projects of clear local and even private interest (Shallat 1994: 127–40; Larson 2001: 141–8; Hill 1957: 170–80). Thus began a contest between rational administration and congressional politics that remains unresolved and contentious into the twenty-first century.

Instead of national planning, Congress settled on a piecemeal approach to public works development. About three weeks after passage of the General Survey Act, President Monroe signed legislation appropriating $75,000 to improve navigation on the Ohio and Mississippi rivers—major routes to the western part of the country. The act empowered him to employ 'any of the engineers in the public service which he may deem proper' and to purchase the 'requisite water craft, machinery, implements, and force' to eliminate various obstructions. The two acts together initiated the permanent involve-ment of the Army Corps of Engineers in rivers and harbours work. How-ever, each act focused on one activity: the General Survey Act on planning; the Ohio-Mississippi legislation on construction. Two years later, Congress combined surveys and projects in one act, thus establishing a pattern that lasts until the present. The 1826 act, therefore, can be called the first true rivers and harbours legislation.

By the time of the Civil War, the federal contribution to river, harbour, and canal improvements amounted to about $17 million in appropriated monies. Some 4.6m. acres of public lands were given for canal improve-ments and another 1.7m. acres for river improvements. Land grants under the 1849 and 1850 Swampland Acts and the 1841 Land Grant Act totalled some 73 million acres. While these grants and appropriations were signifi-cant, they represented a modest amount of aid compared with state and private sector contributions, which by 1860 totalled well over $185m. for canals alone (Reuss 1991: 5). Corporations and public agencies spent many millions more on the construction of urban water systems.

Many of the canal companies incorporated by the states ran into trouble. The 1837 depression had driven a number into bankruptcy; others survived, but only with a healthy influx of state money, state-guaranteed bonds, and occasional federal and state land grants. Often, too, the national story was repeated at the state level, with rationally planned canal routes sacrificed to local political pressures to extend canals to uneconomical out-of-the-way villages. The one major exception to this sad story was the Erie Canal, whose success had spurred the canal boom that increasingly appeared more like a dismal bust, especially with new competition from the railroads. From Pennsylvania to Ohio to Indiana to Illinois and on into the states of the old Northwest, canal fever turned to canal panic, and the public lost faith in both the companies and the politicians who had supported the enterprises (Larson 2001; Goodrich 1960).

The American Civil War (1861–5) also affected the development of water projects. Military action and wartime budgetary constraints took their toll on many of the nation's ports and navigable waterways, and after the war commercial development accelerated demands for waterway improvements. A business-oriented Republican Congress responded by authorizing a great deal more money for rivers and harbours. The federal government also took over many of the bankrupt canal companies, and the Corps of Engineers became the custodian of former private or state waterways. This, as one author put it, was the 'Golden Age of the Pork Barrel' (Pross 1938). Between 1866 and 1882, the presidents signed sixteen Rivers and Harbors Acts. The 1866 Act appropriated $3.7m. for forty-nine projects and has been described as the first omnibus legislation, so called because, like a horse-drawn omnibus of the time, it provided room for a great many people boosting various projects. Sixteen years later, though, the 1882 Act appropriated five times more money. By that year, the federal government had spent over $111m. on rivers and harbours projects (ibid. 44, 52–3; Johnson 2000). 'Willingness to pay'—the primary test of project implementation before the Civil War—now included unprecedented federal largesse. In the so-called 'Gilded Age', lack of federal or non-federal funds was about the only thing that prevented construction.

By the 1880s the basic working relationship between Congress and the Army Corps of Engineers was set. Congress directed the Corps to survey potential projects, make recommendations, and provide cost estimates. Rivers and harbours acts funded both the surveys and the projects that Congress chose to authorize. Also in the early 1880s Congress mandated that the Corps of Engineers use more contractors and less hired labour. By the end of the century, contractors did nine-tenths of all waterways construction, and no Corps officer could use hired labour without the express authority of the Chief of Engineers (Johnson 2000). Increasingly, then, the Corps became a funding conduit to the private sector. This pattern did not stop private sector engineers from calling for the complete elimination of the Corps from public works, but Congress rejected all bills that leaned in that direction.

Fear of railroad competition and questions about federal aid to projects of apparently local benefit moved the Senate in 1872 to create a Select Committee on Transportation Routes to the Seaboard. Composed of nine senators, the committee was headed by Senator William Windom of Minnesota and known popularly as the Windom Committee. Its 1873 report promoted waterway over railway transportation wherever waterways were properly located. Of more relevance here is the committee's conclusion (on a 5 to 4 vote) that the sum of local rivers and harbours projects contributed to the national interest (US Senate 1874). Generally accepted by Congress, this conclusion justified federal contributions for waterway improvements. The result was the authorization of dozens of dubious projects. By 1907, the cumulative total for rivers and harbours appropriations was more than four times the 1882 figure (and, of course, even greater if inflation is taken into account); the federal role in navigation improvements continued to grow.

Meanwhile, the issue of constitutional authority had somewhat changed focus. In 1870, the Supreme Court ruled in *The Daniel Ball* case that the common-law doctrine that navigability depended on tidal influence, a doctrine accepted in British courts, did not fit the American situation. However, the definition the Court substituted was extraordinary. The test of navigation was to be the river's 'navigable capacity'. That meant, the Court went on:

Those waters must be regarded as public navigable rivers in law which are navigable in fact. And they are navigable in fact when they are used, or are susceptible of being used, in their ordinary condition, as highways for commerce over which trade and travel are or may be conducted in the customary modes of trade and travel on water. (Quoted in Hoyt and Langbein 1955: 166)

In short, American rivers were navigable if they were, are, or could be navigable. This decision, in combination with the earlier 1824 Court ruling, made the federal government the clear guardian and ultimate decision-maker on tens of thousands of miles of waterways in the United States. In practice it sufficed to show that a stream had the capacity to float logs to declare it navigable (Smith 1909: 33). However, with this issue more or less settled, another appeared: flood control.

Rivers always flood, but the floods do not always damage life and property. In the United States, we can trace floods as far back as 1543, when

Mississippi River floods stopped Hernando De Soto's expedition (Hoyt and Langbein 1955: 335). Naturally, as settlers moved into the floodplains and built villages, then cities, the damages increased. By the mid-nineteenth century, the problem was becoming critical along the lower Mississippi. Most people put their faith in technology to protect them. Indeed, the then popular term 'flood prevention' testifies to an extraordinarily unrealistic idea when one thinks about it. In the twentieth century, the term became 'flood control', a somewhat more modest formulation. Nowadays we speak of 'flood damage reduction', which probably comes closest to the mark. In any case, in the 1870s calls came for repairing and raising the levees on the Mississippi River. In 1879, Congress created a joint military–civilian Mississippi River Commission to develop and implement plans to improve navigation and flood control on the lower Mississippi. However, once again some Congressmen raised constitutional objections, expressing doubts that flood control was an appropriate federal activity. Until 1890 no appropriation could be used for repairing or constructing any levee in order to prevent damage to lands from overflow, or for any purpose other than deepening and improving the navigation channel (ibid. 172). The 1890 floods along the lower Mississippi resulted in the removal of this restriction, which, in any event, had had little practical effect other than satisfying congressional scruples.

Floods in 1912, 1913, and 1916 along the Ohio and Mississippi rivers eventually led to passage of the 1917 Flood Control Act, the nation's first act dedicated solely to flood control. It provided funds on a cost-shared basis for levee construction along the lower Mississippi and another appropriation to improve the Sacramento River in California. While an important step towards federal involvement in flood control, it was comparatively modest compared to what followed in the coming decades.

Congressional involvement in flood control happened incrementally and timorously. In comparison, its involvement in reclamation projects came quickly. Although Congress had appropriated money for various Western surveys and studies dating back to the early nineteenth century and had also passed several pieces of legislation making land 'dirt cheap'—no other expression suffices—for those willing to cultivate it, it had not seriously considered government assistance to develop Western water potential until the beginning of the twentieth century. As with flood control, some Congressmen raised constitutional objections, questioning the federal authority for condemning water in one state for use in another. The 1902 Reclamation Act created a revolving fund in which all the proceeds from public land sales were placed in the hands of the Secretary of the Interior to be used to construct irrigation works in the West. By law, more than half the revenue from land sales in a state was to be expended within that same state (Pisani 1992: 273–325). Eventually, the fund provided for hundreds of federal irrigation projects that today dot the Western landscape, including such engineering marvels as Hoover and Glen Canyon dams on the Colorado River, the Central Valley Project in California, Pathfinder Dam in Wyoming, and Arrowrock Dam in Idaho. Nor can we forget the canals, pumps, and irrigation channels supported by the revolving fund. They may be less dramatic than the huge dams confined between looming canyon walls, but they are just as essential for irrigation. In theory, annual payments from project users would replenish the fund.

Initially, the reclamation fund totalled $6 million. The time limit for project repayment was set at ten years, then extended to twenty years in 1914, and to forty years in 1926. No interest charges were assessed. The law specified that water would not be provided to any tract of more than 160 acres, and it authorized the Secretary of the Interior to withdraw from the public domain any land necessary for project development (Dickerman Radosevich, and Nobe 1970). Homesteaders filed free claims to the land and received title to it after five years of residence. They quickly found ways to circumvent the 160-acre limitation, using partners and family members to gain control of a much larger tract. In fact, the law fostered a speculative frenzy. Many simply grabbed the free land and, once water was available, sold plots at inflated prices to true homesteaders, thereby forcing latecomers to start heavily in debt. The act also established a Reclamation Service within the US Geological Survey. In 1907 the Service became a separate agency within the Department of the Interior, and in 1923 the name was changed to the Bureau of Reclamation (Worster 1985: 170–1). The authorizing legislation, as amended, confined the Bureau's responsibility to seventeen Western states.

At the beginning of the twentieth century, Ohio Representative Theodore Burton, chairman of the House Rivers and Harbors Committee, challenged the so-called congressional 'pork barrel'. He believed that if non-federal interests—states and communities—partially funded projects, marginal projects would be weeded out. He also successfully promoted in 1902 the establishment of a Board of Engineers for Rivers and Harbours within the Corps of Engineers to review the cost-effectiveness and feasibility of rivers and harbours projects recommended by lower-level engineer officers. However, he opposed Progressive Era conservation proposals that, like some earlier ideas, would grant more power to the executive branch, usually through the creation of a board to plan and approve multipurpose projects that addressed a wide variety of needs, including navigation, flood control, irrigation, water supply, and hydropower. Multipurpose advocates thought the less water 'wasted', the better. Rational, scientific management would replace crude political calculations. Scientific efficiency rather than 'willingness to pay', would guide the planning and construction of water projects (Reuss 1991: 7).

President Theodore Roosevelt embraced multipurpose planning completely. He appointed an Inland Waterways Commission, composed of four government experts, two senators, and two representatives, to propose a comprehensive multipurpose plan for water development. Senator Francis G. Newlands of Nevada proposed yet another commission to carry out the plan. Newlands' proposal was Burton's worst fear. This new executive branch commission of experts would oversee the water programme and could withdraw funds from an Inland Waterway Fund without further congressional authorization. A majority in Congress, and just about every army engineer, shared Burton's concern, partly because of fear of executive branch growth and partly because the bill threatened Corps domination of federal water projects (Hays 1959: 105–10). Burton supported a substitute bill specifying that the commission would act only 'as authorized by Congress' (quoted ibid. 113). In 1908, the House overwhelmingly passed the bill, but the Senate killed it. The 1917 Rivers and Harbors Act actually authorized a waterways commission composed of seven presidential appointees. But President Woodrow Wilson never made any appointments, and Newlands' death in 1919 eliminated the act's major champion. In 1920, Congress repealed the waterways commission and instead established a Federal Power Commission.

Some army engineers objected to multipurpose projects because of constitutional reservations. More raised technical concerns over multipurpose reservoir operations. It was not clear, after all, how to operate a reservoir to respond to both hydropower, which requires a relatively full lake, and to flood control, which requires that the reservoir be as empty as possible to accommodate upstream floodwater. How would the engineers hold back water for later release to aid navigation as well as release the water to meet irrigation, water supply, and hydropower demands? The difficulties were many, and they remain so. None of this, however, impeded the Corps' performance when Congress gave it the responsibility in 1927 to prepare general multipurpose plans to improve navigation, waterpower, flood control, and irrigation for all the navigable rivers of the United States that seemed capable of supporting hydropower. The resulting '308 reports', named after the House document in which the cost estimates for the reports first appeared, provided basic data for multipurpose development for decades to come (Reuss 1992: 106).

The most successful co-ordinated efforts at water control responded to common economic requirements that transcended state borders. These requirements became pressing at the beginning of the twentieth century as a result of two unrelated developments: the need for irrigation water in the West and the growing demand for electrical energy throughout the country. The first development called for institutional, technological, and legal arrangements to allocate scarce water supplies throughout the West. The second called for harnessing the nation's rivers to produce hydropower. The two developments coalesced in 1922, when the states in the Colorado River basin (except Arizona, which joined in 1929) signed the Colorado River Compact. Congress ratified the compact in December 1928 and also authorized the building of a great multipurpose dam in the Black Canyon of the Colorado: Boulder (Hoover) Dam. This initiated the era of regional compacts designed to make efficient use of the nation's rivers. Generally, these regional arrangements mirrored hardheaded political realities more than farsighted planning. When Boulder Dam was authorized, few anticipated a string of dams stretching from the Rocky Mountains nearly to the Mexican border.

Also in 1928, following a devastating flood the previous year, Congress authorized a massive flood control plan for the lower Mississippi River that substantially enlarged federal responsibility for the Mississippi beyond that provided in the 1917 Flood Control Act. The 1928 Act authorized the Army Corps of Engineers to build levees and revetments, dredge rivers, construct outlets, and formulate plans for flood protection for the entire lower Mississippi Valley. Except for the donation of rights-of-way for tributary levees and floodways, the project was to be built at full federal cost. This was both a technological and political experiment. Here there was no interstate compact to regulate water use, and no formal state approval was required. While the federal government's right to regulate interstate navigation had long been generally recognized, the 1928 Flood Control Act significantly expanded the national government's involvement in planning, implementing, and managing interstate flood control projects (Reuss 1998).

In the New Deal of Franklin Delano Roosevelt, river basin planning became a social experiment, and the Tennessee Valley Authority— developer of an area four-fifths the size of England—became the prototype. Questions abounded. Did the TVA administer a cultural, geographic, or natural resource region? What objectives should the TVA have and would they threaten traditional institutions and patterns of life? Were the engineering solutions economically efficient and socially beneficial throughout the basin, and did they address both short- and long-term needs? The TVA became a social laboratory, and, while it successfully provided electricity to the region, some of the social experiments initially envisioned were never implemented (Reuss 1992).

The Corps of Engineers began calculating benefits in the early twentieth century, but it was only in the 1936 Flood Control Act, which established flood control as a proper nationwide federal function, that Congress formally required benefit–cost ratios (Porter 1995: 148–89) The Act specified that benefits 'to whomsoever they accrue' should be ascertained, a requirement that enabled planners to consider an area much larger (or smaller) than the watershed to justify multipurpose development. The act also specified that benefits must exceed costs before projects could be constructed. In the following decades, various interagency committees and the Bureau of the Budget developed criteria based on classical welfare economics to try to optimize net benefits. Instead of scientific efficiency, which had emphasized maximum water development, planners pursued economic efficiency. They looked at regional and national costs and benefits, including traditional objectives such as reducing flood damages as well as new concerns such as preserving ethnic enclaves, and, increasingly, reducing impacts on the environment.

The impact of the 1936 Flood Control Act on subsequent federal water resources development can hardly be overestimated. The legislation authorized 211 flood control projects—principally levees, reservoirs, and drainage channels—in thirty one states at an estimated cost of approximately $300m. (Hoyt and Langbein 1955: 175; Arnold 1988). Congress passed it in response to the suffering and devastation caused by the spring floods of 1936 and also to alleviate unemployment during the Great Depression. In the absence of floods and economic depression, it is doubtful the legislation would have reached the President's desk. Although the Act authorized only single-purpose flood control projects, most of the reservoirs authorized ultimately became multipurpose. The Act specified that non-federal interests contribute the lands, easements, and rights-of-way, hold the government free from damages due to the project, and operate and maintain the works. In 1938 Congress passed legislation that effectively eliminated these requirements for flood control dams and reservoirs and for channel improvement projects (49 Stat. 1570) but three years later restored them for channel and local protection projects (85 Stat. 638). The federal government continued to assume the full cost of constructing and maintaining navigation projects and flood control dams.

To those who still had reservations about the constitutionality of flood control, the United States Supreme Court supplied a definitive answer in 1940 in *United States* v. *Appalachian Electric Power Company*. In that decision, the Court ruled that flood control and watershed development come under the Commerce Clause of the Constitution. The following year, the Court pointed out in *Oklahoma* v. *Atkinson*, 'There is no constitutional reason why Congress cannot, under the commerce power, treat the watersheds as a key to flood control on navigable streams and their tributaries . . . there is no constitutional reason why Congress or the courts should be blind to the engineering prospects of protecting the nation's arteries of commerce through control of the watersheds' (quoted in Hoyt and Langbein 1955: 166–7). In a case before the Court in 1950 (*United States* v. *Gerlach Live Stock Co.*), the justices ruled that 'large scale projects for reclamation, irrigation, and other internal improvements' also fell under the constitutional provision to provide for the general welfare (quoted ibid. 167–8). Thus, constitutional questions were effectively laid to rest on these issues after more than 150 years of ambiguity and acrimony.

The great dam-building era in American history followed passage of the 1936 Flood Control Act. Construction of Hoover Dam on the Colorado (the largest in the world upon completion), Bonneville and Grand Coulee dams on the Columbia, Fort Peck dam on the Missouri, the Bureau of Reclamation Central Valley Project in California, and several other dam projects had already commenced prior to passage of the Act. Fort Peck, Grand Coulee, and Bonneville had been started with emergency appropriations funds at the direction of President Roosevelt in response to the need for unemployment relief during the Depression. Among other projects, the 1936 Act authorized the Los Angeles Flood Control System, dams in New England, and a system of dams in the upper Ohio River valley. Subsequent amendments in the next ten years authorized a system of large dams along the Missouri River and more dams on the Columbia. Meanwhile, the Corps also constructed a system of locks and dams on the upper Mississippi River. The Bureau of Reclamation built Shasta Dam in California and numerous dams on the Colorado, culminating in Glen Canyon Dam, completed in the mid–1960s.

All these projects, numbering eventually into the hundreds, signified a major shift in the federal contribution to water projects, as can be seen from figures provided by the Hoover Commission on the Organization of the Executive Branch of the Government in 1955. Table 3.1 shows the federal investment in water resources broken down into chronological periods.

The Commission made no attempt to convert the numbers to current dollar values. Most of this federal investment, as we have seen, was in navigation and flood control projects. Irrigation and hydropower remained largely in the hands of private, local, or state entities. Federal investment in hydropower had increased from 1 per cent of the total in 1930 to over 13 per

TABLE 3.1. *Federal investment in water resouces, 1824–1954*

Period	Expenditure ($bn.)	%
1824–1920	1.15	8
1920–1930	0.86	6
1930–1945	2.58	18
1945–1954	9.73	68
TOTAL	14.32	100

Source: Reuss and Walker 1983: 1.

cent by 1953, making the federal government (mainly TVA, the Bureau of Reclamation, and the Corps of Engineers) the largest producer of hydropower in the country (ibid.).

In 1968, Congress established the National Water Commission to assess the country's water needs and to recommend improvements in both the planning and construction of projects. The Commission's 1973 report converted all figures to the 1972 dollar value and came up with estimated contributions (see Table 3.2) for water projects to the end 1969 (ibid. 3).

Comparing the Hoover and National Water Commission reports, we see an exponential percentage increase in federal contributions in the 27 years following World War II.

If we translate some of these funding figures into dam projects, we obtain an even more revealing perspective. The approximate number of dams built in the United States and still standing in 2001 is shown in Table 3.3 (several

TABLE 3.2. *National Water Commission estimate of conributions to water projects to end of 1969*

Contributor	Estimated amount ($bn.)
State and local interests	194.5
(of which $180bn. for municipal water and sewage treatment)	
Private interests	56.5
Federal	87.7
TOTAL	338.7

Source: Reuss and Walker 1983: 3.

TABLE 3.3. *Dams built in the United States and still extant in 2001*

Date of construction	No. still standing
Before 1900	2,532
1900–19	4,034
1920–39	5,968
1940–59	15,441
1960–69	19,310
1970–79	13,076
1980–89	5,017
1990–2001	2,557
TOTAL	67,935

thousand dams are not included because their dates of construction are not known).

Federal agencies and independent offices and commissions owned about 4,000 dams. The construction of federal flood control dams peaked in the 1960s, with the number of Corps of Engineers flood control facilities growing at an average annual rate of 6 per cent. The Corps completed 95 major flood control dams during the decade, while the Soil Conservation placed over 2,000 small watershed dams into service. Since that decade, the pace of construction has declined. Quite clearly, the golden age of dam construction, both federal and non-federal, occurred during the immediate post-World War II period (Federal Emergency Management Agency 2001: 36; Federal Interagency Floodplain Management Task Force 1992: 12. 9–13; Pearre 2001–2).

Post-World War II federal dam construction came despite numerous presidential efforts to reduce the federal largesse. President Eisenhower stressed local responsibility and tried to decrease strains on the federal budget by eliminating uneconomical or otherwise undesirable projects. He particularly wished to limit the federal role in waterpower development and to confine federal assistance under the small watershed programme of the Soil Conservation Service. President Carter proposed a 'hit list' of uneconomical or environmentally damaging projects, but in the end capitulated to Congress on many of them. President Reagan came into office with a programme that emphasized reducing the size of government and shifting some of the financial burden to states and communities and the private sector. Instead of attempting to cajole Congress into limiting water projects, as Carter had, Reagan far more successfully changed policy through budget manipulation. 'Budget is policy', was the lesson (Bartlett 1984: 121).

While recent presidents have periodically expressed dismay over wasteful water projects, it was congressional fragmentation rather than Executive Branch opposition that led to years of famine for water developers. More demands on the federal budget in the 1970s and 1980s meant that discretionary programmes such as water resources became candidates for fiscal restraint. Water projects amounted to only half of 1 per cent of the federal budget but to a little over 3 per cent of the discretionary budget subject to the budgetary axe (Reuss 1991: 65–6). In consequence, water interests fragmented, fighting among themselves for a decreasing share of the federal pie rather than mobilizing a strong, united front as they once had been able to do. Meanwhile, environmental organizations increased in strength and challenged some of the congressional pet projects. Changes in the congressional seniority system meant that some of the Corps' long-time supporters in the House and Senate no longer enjoyed the clout they once had. Nor in the environmental era was Congress apt to receive as much public support as formerly for water projects. All of this meant that between 1970 and 1986, Congress passed no significant authorizing legislation for water resources.

Finally, in 1986 Congress passed and the President signed the Water Resources Development Act of 1986 (WRDA–86). It authorized 377 new projects for construction or study. More important, however, were the policy changes. The act put more of the financial and management burdens on the backs of non-federal interests, firmly integrated environmental considerations into water resources planning, and attempted to establish a process to reduce the number of marginal and uneconomical projects. It authorized about $16bn. in spending for water projects, of which the federal government would pay about $12bn. The Act required non-federal interests to pay 25–35 per cent towards the cost of flood control projects. Since 1978 inland waterway users had been assessed a user's fee, a tax on fuel sales for inland waterways traffic, to offset the costs of constructing and maintaining the vast inland lock and dam system. WRDA–86 confirmed the policy, instituting a programme of incremental increases in the fee over the next several years. Thus, WRDA–86 compelled beneficiaries to help fund water resource benefits, reversing the full federal funding that had supported navigation interests for 200 years and flood control dam beneficiaries for nearly fifty years. However, the philosophy behind these reforms was hardly revolutionary. Indeed, in putting more initiative, as well as the funding burden, in the hands of non-federal interests, the act was profoundly conservative, for it restored the federal–state relationship regarding water development that existed during much of the nineteenth century (Reuss 1991: 1–2).

Today, federal water resource agencies do far less structural development than they did a few decades ago. The 'big dam era' that lasted from the 1930s to about 1980 may well be seen as a blip on the screen in a few years. Practically, only so many dams can be built; the reservoirs behind US dams currently store about 60 per cent of the entire average annual river flow of the country (Gleick 1998: 70). Future projects probably will be more closely tied to watershed management and ecosystem restoration. Billions of dollars may be spent to undo what federal water agencies, pursuant to congressional direction, did earlier. The outstanding example is restoring the natural flow of the Kissimmee River, Lake Okeechobee, and the Everglades in south Florida. Congress has authorized a $7.8bn. appropriation for this project, and that will probably not be enough. Beginning in the early 1980s, more money has been spent on maintaining and operating facilities than on constructing new ones, although the Corps has calculated that 3,000 dams in the United States are unsafe and numerous locks on the Ohio, Upper Mississippi, and Columbia rivers are too old, dilapidated, and small to serve modern shipping (Reuss 1991: 38–9). Meanwhile, probably a minimum $100bn. is needed over the next twenty years to modernize water supply and wastewater treatment facilities in the United States (Frederick 1991: 65). The US water resources infrastructure is obviously vital to the country's economy, so there is little question of letting it fall into disrepair. Yet, water planners must take into account both the economic benefits and the environmental costs, while politicians calculate how they can provide necessary services without increasing taxes or mortgaging a community's future through the bond market. There is no easy answer now, just as there was no easy answer 200 years ago.

Not only do we choose different projects, but we plan their design, construction, and, as we have seen, financing in ways that separate today's planning from that of a few decades ago. Indeed, the United States has entered a new era in planning, not formally recognized, but nevertheless manifestly evident. Replacing both the scientific efficiency model of the early twentieth century and the more recent economic efficiency model (which still formally remains) is an approach that compels planning by constraints. The process emphasizes regulation and focuses on water quality issues rather than on quantity. Instead of maximizing economic efficiency or optimizing the opportunity to meet public objectives, it sets limits to growth. Legal constraints include the National Environmental Policy Act (1969), which requires environmental impact statements for any federal project likely significantly to affect the environment, the Federal Water Pollution Control Act Amendments of 1972, and the Endangered Species Act (1973).

One possibly unforeseen impact of environmental legislation, especially of laws touching on water quality or on non-structural flood control projects, is the greater consideration given to concerns of ethnic minorities, the inarticulate, and the poor whose lives and property had often been sacrificed on the altar of national economic development. Issues of environmental justice that fifty years ago were easily ignored are now seriously addressed. Agencies increasingly favour alternative dispute resolution techniques such as arbitration and mediation to respond constructively when impasses threaten. These techniques keep disputes outside the courtroom rather than leave to the judicial system difficult decisions involving questions of equity and ethics.

Finally, we come back to answer the question posed at the beginning of this chapter: if liberty is to be constrained in exchange for access to adequate water of acceptable quality, Americans evidently prefer that it be done at the local, not the national, level Growing technical competence among non-federal entities buttresses this cultural preference. In the last few decades, states and communities have hired their own engineering experts and need not depend so heavily on federal water agencies. Often, federal money is accepted only if the constraints on local decision-making are acceptable. Even Western irrigation reflects this bias in favour of local initiative. The Bureau of Reclamation provides water to only 20 per cent of the irrigated land in the seventeen Western states (Storey 2002). Navigation improvements may be the only water-related area in which Americans still look to the national government for leadership, but today the water transportation industry helps offset the expense of construction and maintenance. In fact, all the elements that framed water resources planning in the early United States are still evident: distrust of government expansion, ambiguous boundaries between state and federal power, constitutional questions relating nowadays to wetlands regulations and the 'taking' of private property, a general deference to the private sector guided by Adam Smith's ubiquitous 'invisible hand', and political sectionalism that defeats rational national planning.

Water resources development will ever test the nature of American republicanism, as the boundaries between state and national power shift and as the border between liberty and authority responds to changing circumstances. Expensive water projects often require cost-sharing, respond to the needs of a large number of economic and social groups, and may affect large regions that embrace multiple jurisdictions and levels of government. Consequently, their planning and construction test the resilience of American institutions and challenge the nation to seek cooperative answers rather than capitulate to a much easier solution: authoritarian direction. It is not too much to say that America's answers to its water resources needs help to form the very contours of its democratic process.

REFERENCES

Albjerg, V. L., 'Internal improvements without a policy (1789–1861)', *Indiana Magazine of History*, 28 (1932), 168–79.

Appleby, J., *Inheriting the Revolution: The First Generation of Americans* (Cambridge, Mass.: Harvard University Press, 2000).

Arnold, J. L., *The Evolution of the 1936 Flood Control Act* (Fort Belvoir, Va.: Office of History, U.S. Army Corps of Engineers, 1988).

Bartlett, R. V., 'The budgetary process and environmental policy', in N. J. Vig and M. E. Kraft (eds.), *Environmental Policy in the 1980s: Reagan's New Agenda* (Washington, DC: CQ Press, 1984), 121–41.

Dickerman, A. R., Radosevich, G. E., and Nobe, K. C., *Foundations of Federal Reclamation Policies: An Historical Review of Changing Goals and Objectives* (Fort Collins, Colo.: Department of Economics, Colorado State University, 1970).

Elazar, D. J., 'Federal–State collaboration in the nineteenth-century United States', in D. J. Elazar *et al.* (eds.), *Cooperation and Conflict: Readings in American Federalism* (Itasca, Ill.: F. E. Peacock, 1969), 83–108.

Federal Emergency Management Agency, *The National Dam Safety Program, Fiscal Years 2000–2001* (Washington, DC: Federal Emergency Management Agency, 2001).

Federal Interagency Floodplain Management Task Force, *Floodplain Management in the United States: An Assessment Report,* ii.: *Full Report* (Washington, DC: Federal Emergency Management Agency, 1992).

Frederick, K. D., 'Water resources: Increasing demand and scarce supplies', in K. D. Frederick and R. A. Sedjo (eds.), *America's Renewable Resources: Historical Trends and Current Challenges.* (Washington, DC: Resources for the Future, 1991), 23–78.

Gallatin, A., *Report of the Secretary of the Treasury on the Subject of Public Roads and Canals Made in Pursuance of a Resolution of Senate of March 2, 1807* (Washington, DC: R. C. Weightman, 1808).

Gleick, P. H., *The World's Water: The Biennial Report on Freshwater Resources, 1998–1999* (Washington, DC: Island, 1998).

Goodrich, C., *Government Promotion of American Canals and Railroads,1800–1890* (New York: Columbia University Press, 1960).

Harrison, R. W., *Alluvial Empire* (Little Rock, Ark.: Delta Fund in Cooperation with the US Department of Agriculture, 1961).

Hays, S. P., *Conservation and the Gospel of Efficiency: The Progressive Conservation Movement, 1890–1920.* (Cambridge, Mass.: Harvard University Press, 1959).

Hibbard, B. H., *A History of the Public Land Policies* (Madison: University of Wisconsin Press, 1965).

Hill, F. G., *Roads, Rails & Waterways: The Army Engineers and Early Transportation* (Norman, Okla.: University of Oklahoma Press, 1957).

Hoyt, W. G., and Langbein, W. B., *Floods* (Princeton: Princeton University Press, 1955).

Johnson, L., 'The Corps in the Gilded Age, 1866–1896', Unpublished manuscript. (Office of History, US Army Corps of Engineers, 2000).

Kelley, Robert, *Battling the Inland Sea: American Political Culture, Public Policy, and the Sacramento Valley, 1850–1986* (Berkeley: University of California Press, 1989).

Kemper, J. P., *Rebellious River* (Boston: Humphries, 1949).

Larson, J. L., *Internal Improvement: National Public Works and the Promise Of Popular Government in the Early United States* (Chapel Hill: University of North Carolina Press, 2001).

Maier, P., 'The revolutionary origins of the American corporation', *The William and Mary Quarterly,* series 3, 1 (1993), 51–84.

Pearre, C. M., telephone and e-mail communications from Mr. Pearre, Engineering and Construction Division, Headquarters, US Army Corps of Engineers, Washington, DC, (2001–2).

Pisani, D., *To Reclaim A Divided West: Water Law, and Public Policy, 1848–1902* (Albuquerque: University of New Mexico Press, 1992).

Porter, T. M., *Trust in Numbers: The Pursuit of Objectivity in Science and Public Life* (Princeton: Princeton University Press, 1995).

Pross, E. L., 'A History of Rivers and Harbors Appropriation Bills, 1866–1933', Unpublished Ph.D. dissertation. (Ohio State University, 1938).

Reuss, M., *Designing the Bayous: The Control of Water in the Atchafalaya Basin, 1800–1995* (Alexandria, Va.: Office of History, US Army Corps of Engineers, 1998).

—— 'Coping with Uncertainty: Social Scientists, Engineers, and Federal Water Planning', *Natural Resources Journal*, 32 (1992), 101–35.

—— *Reshaping National Water Politics: The Emergence of the Water Resources Development Act of 1986* (Alexandria, Va.: Institute for Water Resources, US Army Corps of Engineers, 1991).

—— Walker, P. K., *Financing Water Resources Development: A Brief History* (Washington, DC: Historical Division, Office of the Chief of Engineers, 1983).

Scheiber, H. N., *The Condition of American Federalism: An Historian's View*, Subcommittee on Intergovernmental Relations of the Committee on Government Operations, United States Senate, 89th Cong., 2nd Sess. (Washington, DC: US Government Printing Office, 1966).

Shallat, T., *Structures in the Stream: Water Science, and the Rise of the U. S. Army Corps of Engineers* (Austin: University of Texas Press, 1994).

Smith, H. K., United States Commissioner of Corporations, *Report of Commissioner of Corporations on Transportation by Water in United States*, (Washington, DC: US Government Printing Office, 1909).

Storey, B., telephone communications from Dr Storey, Senior Historian, US Bureau of Reclamation, Denver, Colo., 2002.

United States Senate, *Select Committee on Transportation-Routes to the Seaboard*, 43rd Cong., 1st Sess. (Washington, DC: US Government Printing Office, 1874).

Wills, G., *A Necessary Evil: A History of American Distrust of Government* (New York: Simon & Schuster, 1999).

Wilson, W., 'The Study of Administration', *Political Science Quarterly*, 2 (1887), 197–222.

Wood, G. S., *The Creation of the American Republic, 1776–1787* (Chapel Hill: University of North Carolina Press, 1969).

Worster, D., *Rivers of Empire: Water, Aridity, and the Growth of the American West* (New York: Random House, 1985).

4

Managing Water for the Future:
The Case of England and Wales

Ian Byatt

To understand how things have worked is the best preparation for looking ahead. So I will not gaze into a crystal ball but explain what is happening in England and Wales. I will, however, set out and discuss some scenarios for the future.

THE PROGRESS OF THINKING

In 1989 the water industry emerged from the nationalization era which it had entered only fifteen years earlier. It was a late entrant into the world of public corporations that had emerged between the wars, and particularly after 1945—a world that was a product of Fabian thinking and wartime experience. The Fabians provided the intellectual base for 'gas and water socialism' in the late nineteenth and early twentieth centuries. Two world wars encouraged people to believe that the state could manage our basic industries efficiently, and the inter-war depression drew attention to deficiencies in the working of the market economy.

'Gas and water socialism' started in the last quarter of the nineteenth century, in the municipalities, with gas, water, electricity, and tramways. In the inter-war years there was a movement towards regional, then national operations, culminating in the post-war Nationalization Acts. Consolidation in water followed slowly. The amalgamation of municipal undertakings into ten Regional Water Authorities did not take place until 1973. It brought a host of water and wastewater undertakings onto a river basin basis. A further step was taken in 1983 with the substitution of smaller, more executive boards for the much larger bodies that had included local authority representatives.

THE NATIONALIZED WORLD

The model for nationalization in the UK developed from the experience of Herbert Morrison, a key figure in the post-war Labour Government. It involved an arm's-length relationship with government. By the 1970s, the flaws in this model were evident.

The boards of the nationalized industries were required to act in the social interest, subject to breaking even financially. The definition of the social interest was the responsibility of the boards, without any clear mechanisms for ministers to influence their decisions. It was never clear what 'breaking-even', 'taking one year with the next', meant in practice. Moreover, having delegated social functions to such a public not-for-profit body, ministers found it difficult to stay clear. The power of general direction of the nationalization statutes was never used but there was regular and frequent non-statutory involvement in the affairs of the nationalized industries.

When I worked in the Treasury in the 1970s, there were constant disagreements between the boards of the industries and ministers and, within government, a gulf between those who were concerned with 'the efficient allocation of resources' (Treasury speak) and 'social and sectoral considerations' (Departmental speak). The Treasury was concerned with the scale, and poor quality, of their investment programmes and about their frequent loss-making activities. The Sponsor (*sic*) Departments were concerned with sectoral strategies, such as fuel policies, and social objectives, such as not raising prices for essentials of life. There was also a running concern with nationalized industry pay, because of its effect on pay in the economy generally. Pay policy inevitably involved interference with commercial objectives and behind-the-scenes deals.

The nationalized industries represented the high point of producer domination of business decisions. They provided what *they* thought the nation required. Until 1978 there were no performance objectives or service standards. There was no ultimate sanction for failure to meet performance standards or efficiency (or cost-reduction) targets. Consumer committees were established but never given a significant role. Operational matters escalated to the top as unproductive arguments between senior managers and civil servants—or worse between chairmen and ministers. Investment was constrained and capital was wasted. Industries wanted to invest in grandiose schemes, while the Treasury was concerned to limit the macroeconomic consequences of capital expenditure. Industries complained of lack of investment. Yet financial returns consistently fell short of the cost of capital (Treasury 1961, 1967, 1978).

The institutional structure had produced the wrong incentives.

THE NEW APPROACH: 'GAS AND WATER CAPITALISM'

Conservative ministers elected in 1979 dramatically changed the levers of macro-and microeconomic policy. They resolved to control inflation by monetary policy and to increase labour market flexibility to generate full employment. They believed that competition would liberate consumer choice and put pressures on management to increase efficiency. Traditionally governments had sought to correct market failure. The new approach took account of public sector failure and the power of markets to deliver what customers wanted. This was a profound intellectual change, much approved of by those of us who had observed government failure in our daily work.

Ministers could not remedy investment deficits in the public corporations through the public finances. Privatization was as necessary for investment by BT as it was for investment in the water industry.

'Gas and water capitalism' is not the same as market capitalism generally. The utilities belong to the infrastructure sector of the economy, where there is considerable natural monopoly power through the provision of essential facilities. Independent specialist regulators were, therefore, given statutory powers to act as servants of government in its wider sense, not as servants of ministers. The primary duty of the water regulator was to 'secure that the companies properly carry out their functions and can finance them'. It is for the regulator, not ministers, to decide what constitutes the proper carrying-out of functions. The regulator may consult ministers, but retains final responsibility. The regulator must determine the price limits need to finance the operations and investment of efficient companies, but not how much they should spend.

Subject to these primary duties, the regulator is charged to protect customers, promote efficiency, and facilitate competition. The government wisely provided for appeal by a company to the Competition Commission should it disagree with the regulator's pricing decisions. The regulator appoints ten Customer Service Committees (CSCs) to represent customers. He brought together the chairmen in the OFWAT National Customer Council (ONCC).

Because of the health and environmental importance of water supply and wastewater disposal, the government also appointed specialist quality regulators—a Drinking Water Inspectorate (DWI) and a National Rivers Authority (NRA), subsequently the Environment Agency (EA). Each organization had a specific set of statutory duties. The endemic confusions of the nationalized regime were avoided.

Initially it was not thought that competition was likely to develop rapidly, either in water supply or wastewater disposal. Nevertheless, the Privatization

Act made a start by enabling the regulator to make 'inset' appointments of new suppliers to serve new customers on green field sites within the area of existing companies. In 1992 this was widened to include large customers already served by an existing company.

'GAS AND WATER CAPITALISM': THE RESULTS

Since the privatization of the water authorities in England and Wales in 1989 (OFWAT 2003), water quality has improved:

> The percentage of river and canal water classified as good or fair has risen from 84 to 94 per cent.
> Bathing water compliance has risen from 66 (1988) to 99 per cent.
> Sewage treatment works compliance has risen from 90 to 99 per cent.
> Sewer flooding incidents have fallen from 0.05 (1993) of connections to 0.02 per cent.
> Compliance with drinking water standards rose from 99 to 99.9 per cent.

Customer service has also improved:

> Properties at risk of low pressure have fallen from 1.3 (1993) to 0.06 per cent.
> Unplanned interruptions over 12 hours have fallen from 0.42 per cent of properties to 0.05 per cent.
> Written complaints from customers answered within ten working days has risen from 82 to 99.8 per cent.
> Billing contacts dealt with within five working days has risen from 80 (1992–3) to 99 per cent.

This required a large increase in investment. Investment, running at about £1.5 billion a year (today's prices) in the 1980s, rose to average of £3.0 billion a year during the 1990s. By 2005, some £50 billion will have been invested in the privatized transmission and treatment of water and wastewater.

Customers' bills have had to rise by some 20 per cent about inflation to finance this. Within this increase, however, bills have *fallen* for customers (most business customers and, now, over 20 per cent of household customers) taking a measured supply. Prices would have risen much more but for the greater efficiency of the water companies. The annual average regulated household water bill (now around £230 a year) would have had to rise—between 1990 and 2005—by £100 in order to finance the provision of better quality water and wastewater. In the event it will have risen by £38 (Byatt 2001). Most of the increases in price took place in the first half of the 1990s. Greater efficiency, delivered largely in the second half of the 1990s, enabled the regulator to reduce prices substantially in 2000.

THE FRAMEWORK FOR THE FUTURE

Good governance involves some unbundling of government. Ministers cannot, and should not, try to run everything. Conservative ministers recognized the advantages of distancing themselves from the operations of the utilities. New Labour is taking time fully to learn this lesson. The principles, set out in the Green Paper on Utility Regulation extol the advantages of an arm's-length relationship between ministers and regulators (DTI 1998). Those principles did not, however, prevent the Deputy Prime Minister from wanting to set leakage targets. Nor did they prevent DETR ministers from legislating to give themselves powers—which went beyond the ministerial powers under nationalization—to set water charges for individual groups of customers, and to deny companies the right to disconnect household customers for non-payment of bills.

Politicians are experienced and skilled in playing the zero-sum game of distribution. But they are pressured into action by the politics of the saloon bar. They mistrust the invisible hand of the market and want to pull their own levers—without fully considering all the consequences. And, because they are often fighting battles on several fronts, they baulk at tough economic action, even when they know they should be taking it.

Regulators are perhaps better able to engage in the positive-sum game of enrichment. They need to encourage as well as to complain if things are to be put right; and they need to be patient. They might need to act as a buffer, so that energetic ministers do not overreact to every adverse event. Regulators can also make markets work better: for example, in utility pricing, they can act as a proxy for the market.

EXPOSURE TO THE MARKETS

Before privatization, water was largely insulated from the market economy. Water companies are now part of global business. Expertise, in economics and customer service as well as in engineering, is now freely traded internationally. Water companies raise their own finance and account to their shareholders. They have incentives to increase their profitability by reducing their costs. They are no longer enmeshed in the culture of the public sector labour supply. Customers taking measured supplies now face tariffs that are increasingly related to costs and so have incentives to use water wisely. Businesses have stronger incentives to treat wastewater before discharging it to sewer.

Privatization of water has not simply substituted a private monopoly for a public monopoly. Water companies can use, and are now subject to, market forces in a number of overlapping ways.

1. *Markets for Water Resources.* The abstraction of water from rivers or aquifers is regulated in the interest of the environment. Abstraction licences can, however, be traded, albeit with some complexity. This could usefully be simplified. Water can be bought and sold between companies—with the regulator available to determine the price if the parties disagree. Companies can tap capital markets to finance investment in water abstraction, storage, and transfer schemes. The regulator can allow for this in setting price limits.

2. *Water Product Markets.* Water pricing is being released from the straitjacket of tariffs based on an extinct property tax (rateable values) and from overreliance on standing charges when supplies are metered. Meters are now routinely installed in new properties and there is a steady switch of existing households to metered supplies. If tariffs reflect the continuing cost of supply— including the capital cost of augmenting facilities—customers can enjoy an economic choice of the volume of water bought. Regulators supplement existing market mechanisms.

If, as a result of regulatory and other pressures, the price of water is equal to the incremental cost of supply, including capital costs, water companies are able to finance additional supplies within existing price limits (OFWAT 1993). Where new households are connected to a company's network, it is able to levy an additional, regulated charge to cover the capital costs of new local infrastructure. Large users (water and wastewater) are able to choose their own supplier, and are increasingly doing so. The threshold of the definition of a large user has been reduced. The potential for competition has already reduced prices for large users.

Cross-border competition is possible for domestic supplies, but the scope is limited because borders do not generally pass through well-populated areas. If common carriage is developed commercially, as in telecom, gas, and electricity, customers will get choice and lower prices. Following the 1998 Competition Act, the market is poised for this. The Act requires companies not to abuse a dominant position. OFWAT requires companies to produce a code for access to their networks, including an access price (OFWAT 2002*a*: 1). Potential entrants can take disputes to the regulator. The quality of drinking water in public networks is subject to the inspection of the DWI, or can be made so subject by the process of making inset appointments (see above) for new suppliers.

I believe the market is now contestable. So far there have been few contestants. Yet ministers legislated in 2004 to restrict competition.

3. *The Labour Market.* At privatization the water companies were able to free themselves from historical rigidities in both pay and conditions. And unions are less powerful when ministers are removed from direct involvement. Companies have been able to change internal arrangements to achieve much more efficient working practices, without threats of strikes.

Companies are also able to unlock executive remuneration to pay salaries better related to other private sector companies and to recruit higher quality staff. Where senior managers have failed, shareholders have acted to replace them. Despite the temporary turbulence of the 'fat cat' accusations, a much better framework for executive pay is now in place.

4. *The Market for Procurement.* Privatization also increased the incentive to efficient buying of goods—both capital equipment and operating materials. Savings could be retained up to the next price review. In the case of capital expenditure, this probably affected specifications for projects more than techniques of tendering. One-well publicized example concerns sewerage on the Isle of Wight. Southern Water's original plan for a number of sewerage treatment works along the coast was changed into one that involved piping sewage to a central works and treating it more cheaply to the enhanced standard required by the EU.

As the 1990s progressed, ideas of contracting out substantial elements of operations began to develop into practical schemes. People were even beginning to talk of a 'virtual' utility with a small core staff, ensuring licence conditions were met and letting managing contracts for operations. Following the last price review this activity has intensified and is a major route for water companies to seek to reduce their costs to beat the regulator's projections. Contracting out, i.e. outsourcing—a feature of many private markets in recent years—allows companies to specialize in what they can do best. It enables individual managements to exploit their comparative advantages. Managing distribution networks requires rather different skills from those needed to compete for customers.

As outsourcing develops, it could provide increasing competition while allowing some concentration of the ownership of elements of transmission and distribution networks. Some merger of networks could be combined with increased contracting out of other activities.

5. *The Market for Finance.* Water companies have been, and still are, engaged in an intensive search for new—cheaper and better—sources of finance and to achieve an optimal balance (equity, bonds, loans, etc.) of existing instruments. In contrast to the complexity of estimating the opportunity cost of capital for nationalized industries, the cost of capital for private utilities has become a market phenomenon.

The initial estimates of the cost of capital at privatization were shown to be too high as companies tapped markets in a variety of ways—nominal bonds, indexed bonds, leasing, loans from the European Investment Bank, etc. In the last two years, water companies have shown that they can reduce their cost of capital further by increasing the gearing of a ring-fenced water and sewerage business. Some companies are now talking about gearing (measured by the debt : debt plus equity ratio) of around 85 per cent. This is significantly more than the 50 per cent assumed by the regulator at the 1999 price review.

This has involved new techniques, such as the formation of a private not-for-profit company (Glas Cymru) able to buy Welsh water at a discount to the regulatory capital value (OFWAT, 2001, and the 'thin equity' approaches of Scottish Power/Southern Water Services and Anglian Water (OFWAT, 2002*c*: 2–4 2002*d* and *e*). The emergence of monoline insurers (who combine insurance on capital value or interest payments with lending) may have reduced the cost of borrowing while imposing greater financial discipline on companies. Because water companies, unlike water authorities, pay the full cost of capital, they have an incentive to economize in its use—working existing assets harder and balancing the cost of replacement investment against its benefits.

In a capital-intensive industry such as water there can be additional constraints that are not fully reflected in conventional calculations of the cost of capital. In particular, companies have needed to sustain a level of interest cover necessary to ensure investment grade rating on their bonds. These additional constraints on borrowing were important at all the three water price reviews. Their importance declined as it became progressively clear that levels of gearing could be raised. But they will not disappear while high investment is needed to meet new water quality and environmental obligations.

6. *The Market for Corporate Control.* Water companies are subject to takeover, although legislation restricts internal concentration to preserve a sufficient number of comparators. The threat of takeover is a continuing spur to better performance. The desire to maintain a good share price led companies to try to outperform the cost assumptions the regulator made at a price review. City carrots have proved to be more effective than Treasury sticks. The takeover mechanism made it possible to replace a poorly performing utility, Hyder, without detriment to customers or the environment.

The regulatory restrictions on internal mergers have ensured that a merger would be accompanied by a direct benefit to customers, usually in the form of lower prices. Under this regime there has been significant consolidation of smaller companies and two large mergers, both involving multinational corporations, which were accompanied by substantial reductions in prices. There are no restrictions on entry from outside the water businesses and from outside England and Wales. There have been takeovers by French, German, Spanish, Dutch, American, and Scottish companies.

Markets have shown themselves better able than political processes to cope with institutional change. Incremental changes can be made and tested. The political process struggles with institutional change—too often risking the preservation of obsolete structures or making dramatic changes with unintended effects.

THE REGULATORY COUNTERPART

Incentive regulation is the natural counterpart to the greater use of markets. It has translated greater efficiency into better service and lower prices to customers and into a better environment.

1. *Ring-fencing.* The regulated business must be ring-fenced so that losses (or profits) made by other parts of the company are not attributed to the customers of the regulated business. This involves proper accounting separation, regular visits to understand what is happening, and policing of what companies are doing. It also involves ensuring that any losses incurred in a non-water activity could not impact adversely on the customers of the regulated water utility. This no-recourse rule meant that the collapse of the Enron Corporation, and the downgrading of its bonds to junk status, did not affect the status of the bonds of its subsidiary, Wessex Water, which retained investment-grade status.

The ring-fencing arrangements also involved ensuring that licensed water companies had good governance arrangements to ensure they made decisions in the interests of the water company rather than the diversified parent. There is growing evidence that this is reducing the cost of capital by insulating licensed companies from some of the risks faced by their parents.

2. *Incentives to Efficiency.* The setting of medium-term limits (price caps) on prices charged to customers is a key element in the provision of incentives. OFWAT ensured that companies were able to keep any savings they made on both capital expenditure and operating expenditure for a full five years. So, for example, when Scottish Power took over Southern Water well into the five-year period of the price cap, it was allowed to carry over savings at the next review.

3. *Expectations.* Incentives are not always sufficient. Price limits can—indeed should—be set which require management to work hard to achieve reasonable profits. OFWAT used comparative analysis to establish what expenditure should be allowed for in setting price limits. This involved judging how quickly, and to what extent, companies could be expected to improve their efficiency. Comparative analysis builds on the costs revealed by the more efficient companies, i.e. those who responded best to incentives. Sticks and carrots are linked.

4. *Focus on Outcomes.* Regulation is about achieving outcomes effectively and efficiently, not saving costs by cutting corners. Measures of performance to customers, such as adequate water pressure and continuity of supply, and measures of performance for the environment, such as compliance of sewage treatment work with EU standards, were established and regularly monitored.

The incentives given to water companies to improve customer service standards were strengthened progressively as regulation matured. Initially,

OFWAT recorded, published, commended, and encouraged. Rejecting a call from a Parliamentary Committee to set absolute standards from the regulator's office, we prescribed minimum standards and pushed laggards hard. In the 1999 Price Review, we used a composite indicator of performance. Companies at the top end of this indicator were allowed higher price caps, while those at the bottom were given tougher ones. This is now being developed further, better to incentivize the provision of good service (OFWAT 2002*b*: 5).

5. *Interface with Ministers as Standard Setters.* The independence of the regulator from ministers does not preclude seeking guidance from ministers as setters of water quality and environmental standards. As part of the 1994 periodic review, a set of procedures was established that involved:

1. An open (i.e. public) letter to ministers, setting out the consequences for customers' bills of proposals for improvements in water quality and the environment.
2. Consideration of the issues by ministers collectively and guidance to the water regulator on legally enforceable water quality and environmental obligations.

At the 1994 Review, this included the establishment of an important quadripartite group involving government departments quality regulators (NRA and DWI), companies and OFWAT, under DoE chairmanship. This enabled ministers not to be concerned with the details. Unfortunately, by the 1999 review, ministers had become entangled in detail and the significance of the quadripartite group diminished.

6. *Dealing with Failure.* Monitoring does not always reveal success. Failure has to be dealt with effectively as well as firmly; penalties must be proportional and lead to better results. Failures might simply affect individual customers or could be on a much larger scale. A Guaranteed Standards Scheme (GSS) was set up at privatization so that individual customers would be compensated for specified service failures, such as an unplanned interruption of supply. The CSCs have—in addition—audited complaints and extracted compensation for failures in service. In cases of dispute, companies have ceded binding mediation to CSCs.

The biggest company failure to affect customers occurred in Yorkshire during the drought of 1995. To avoid running out of water, Yorkshire Water took it up the Dales by tanker at a cost of £45m. to the shareholders and loss of sleep to those living by the lorry routes. When the crisis was over, the company commissioned an 'independent' report, which concluded—in those well-worn words—that this was an 'accident waiting to happen' and recommended substantial investment. Ministers wisely kept out of this and left the governmental inquiry to be carried out by the regulator.

The inquiry team was given full access to the company's records. The facts were agreed by the company. The team concluded that there were serious failures in customer service and deficiencies in the governance of the utility business. Also, as then Director-General of Water Services, I concluded that the special dividend of £50m. that Yorkshire Water Services had just paid to its parent company was not appropriate in the circumstances (OFWAT 1996). Following discussions with the new management, the company agreed to lower price limits—costing some £40m. It also agreed formal undertakings to improve service to customers. It agreed to amend its licence to ensure that the utility would conduct its business as though it were a separate plc and not pay dividends which could impair its ability to finance the regulated business.

The episode revealed a hole in the compensation arrangements The legislation apparently did not provide for compensation should the supply be interrupted when a drought order was in place. Despite promises by two administrations to remedy this, nothing has been done. As an 'interim' solution, I negotiated changes in the companies' licences to provide for compensation in all cases. There the matter rests. There is now an automatic financial penalty awaiting a company that fails its supply duty.

7. *Information and Consultation.* Privatization and regulation has led to much greater transparency. Regulators need information and publishing it provides a basis for informed public debate. One of the early jobs in OFWAT was to devise a regular (annual) flow of information from the companies to the regulator. It drew on the internal information systems used by the previous water authorities. It has been modified during the years, in consultation with the companies and other stakeholders.

OFWAT was careful to ensure that information is consistent and comparable. Much midnight oil was burnt in getting the definition right. Independent Reporters were appointed, with a duty of care to the regulator, to validate the data collected by the companies. Wide distribution of this material was ensured, in a series of published annual reports, through *ad hoc* publications, and by making the basic data available to research workers. Key points were set out in leaflets that were widely accessible to customers and to the general public. This information also protects the regulator against asymmetry of information—a very real problem for regulators.

8. *Capital Expenditure.* Much has gone well, but the position on capital expenditure remains imperfect. Investment is not an end result: it is a cost and not an outcome. Only where investment produces a sufficient return does it provide value for money for customers.

Forward estimates by companies are often biased. Investment appraisal is often poor; those promoting schemes can be resistant to the search for better alternatives. Frequently the right information has not been collected. There is a tendency to overspecify work. There is a temptation to press for capital expenditure which can push up price limits in order to reduce operating costs and pay higher dividends. Much regulatory effort has been needed to improve the capital cost estimates used for setting price limits.

By no means all the proposals for investment by water companies have been on worthwhile projects. At the 1999 Periodic Review there were a number of uneconomic projects, notably the expensive Wessex low-flow project, which were postponed for further investigation.

LOOKING AHEAD

The fixed point in looking ahead is the 1999 Periodic Review, which reduced prices in April 2000 by over 12 per cent—the famous (or infamous)—P_0 adjustment. In the first year (2000/1) of the new price limits, prices were reduced so that profits would be no higher than the cost of capital. This contrasted sharply with the approach of the 1994 Price Review where price limits were designed to transfer efficiency to customers by reducing profits to the cost of capital over a ten-year period—although most of the reduction would take place over the first five years.

The logic behind the P_0 (first year) reduction was twofold. First, it was felt that customers, who had endured substantial increases in prices since privatization, should have—and see—the benefit of the efficiency generated by privatization. Secondly, the goal was to ensure transparency in environmental decision-making. Any further environmental—or social—obligations imposed by ministers on water companies would be seen to raise prices. There would be no cushion provided by shareholders.

SCENARIOS FOR THE FUTURE

Attempts to forecast are so often doomed to failure that I have chosen to present my views in the form of three scenarios for the future. In putting them together, I have combined possibilities in a number of variables. The key ones relate to capital expenditure, its financing (i.e. the cost of capital), the structure of the companies, and the nature of the regulatory regime. The broad nature of the pressures on these fronts is clear. The demands for environmental improvements, involving continued capital expenditure on water quality and environmental improvement, are likely to remain strong.

The companies will be much more geared up at the next Periodic Review and the scope for further gearing may be relatively small. The government, meanwhile, is getting increasingly involved in the detailed working of the regulatory arrangements. At the last review, it looked at thousands of environmental improvement projects.

The vertically integrated structure has persisted to a greater extent than in gas and electricity, but is now being challenged as companies increasingly come to concentrate on what they can do best and outsource other activities.

Progress on competition has been slow compared with what has happened in energy. Partly this reflects the underlying situation, partly it reflects the different attitude to competition within government. Environment ministers, particularly under Labour, seem more worried about any downside than excited about the potential benefits. Increased competition is not compatible with detailed control; ministers prefer the control and fear the competition.

Out of this I can image three scenarios: (A) high capex: high prices; (B) competition: high efficiency; (C) muddling through.

Scenario A: High Capex: High Prices. The pressure for a continuation of high capital expenditure could arise from new, poorly targeted water quality and environmental Directives from the EU and poorly appraised national schemes.

By the next price review in 2004 companies may have *projected* interest covers that are barely sufficient to maintain investment grade status on their bonds. If so, there will be pressure on the regulator to advance revenue when setting price limits. Alternatively, appointed water companies may need to raise equity, either through an injection from their parent companies or directly.

It may become increasingly difficult for water companies to raise their efficiency more rapidly than that of the economy as a whole. Many of the gains from the initial transfer from the public to the private sector may already be achieved.

If both capital and operating expenditure stay high, and the cost of capital rises, the price escalator will start again. This will not be popular, whatever the underlying desire for quality. It will exacerbate disputes between those who are content to see bills rising to improve quality and the majority who may not—particularly if other taxes are rising.

Scenario B: Competition: High Efficiency. The effect of competition on energy prices, where large reductions have taken place, shows the scope for increases in efficiency in the water sector once it is driven by competition. Some of this is taking place as companies compete with each other for outsourcing contracts, but the advent of market competition would further stimulate the search for better performance and lower costs.

It is possible substantially to improve investment appraisal, particularly where enhancement projects are concerned. The Environment Agency could, and should, move away from using its political muscle to support any project, however expensive, to developing a system for improving the effectiveness and efficiency of projects, and concentrating on the better ones. The Agency has neglected its statutory duty to consider costs and benefits. It would also improve performance and efficiency if ministers were to concentrate on strategy and not look at a myriad of small schemes—where they and their civil servants have limited expertise.

There is work to be done better to understand capital maintenance, by improving information on service to customers and its relationship to asset conditions and operating costs. When information is available, economic techniques can be used to improve effectiveness and efficiency.

If investment becomes more efficient, costs, including the cost of capital, can stay down. In such a world, we could look forward to steadily improving water and environmental quality, where customers' bills remained broadly stable in real terms.

Scenario C: Muddling Through: Neo-nationalization. In politics, where there are conflicting objectives and no easy answers, muddling through can often be a good strategy. In business life, it is more often the way to oblivion. If water is dominated by politics, we could drift towards neo-nationalization. Ministers would set generalized objectives and intervene in particular events. They would use their extensive powers on water charges to promote social agendas. They would be driven by the sound-bite and the ballot box rather than by business logic.

Nor do ministers have a good track record when it comes to an open assessment of strategic possibilities. The cost of better water and environmental quality was never openly estimated and appraised under nationalization. Immediately after privatization, things even took a turn for the worse. Relieved of having to go to the Treasury for finance for investment, ministers believed the financial markets could be milked. But, as could have been predicted, only at the cost of rapidly rising water prices. It was left to the water regulator to protect customers from environmental taxation, by constantly reminding them about the cost of improving quality.

Muddling through can create tensions while it tries to negotiate through them. Unresolved conflict is the last thing that business leaders want. It leads to paralysis, high costs, and constraints on investment. The end result is poor performance and inefficiency.

WHERE DO WE GO FROM HERE?

It is easier to see who gains from each of the scenarios than to say which will prevail. The companies would welcome Scenario A, the customers, Scenario B. Environmentalists would gain from both, but their instincts probably point them to A. Ministers would instinctively welcome the high investment of Scenario A but recoil from rising prices. Unaided, they may not be able to exercise the self-restraint needed to avoid Scenario C.

My own preference is for Scenario B. It involves recognizing that there are differences in objectives and that centrally imposed solutions do not work. It involves continuing to use markets to allocate resources. It involves unbundling government and involving all stakeholders in making decisions on water and environmental quality. It involves taking risks in encouraging competition, by reducing substantially the threshold for inset appointments, by opening up the market for abstraction of water from rivers and aquifers (subject, of course, to environmental constraints) and by facilitating common carriage, in particular by extending the functions of the DWI to all water supplied through public networks. It involves accepting that capital markets will manage new suppliers. It also involves an evolution of tariffs based increasingly on cost and the extension of metering.

I would scarcely be human if I did not finish this chapter by arguing for maintaining the position of independent regulators, able to act in the public interest at arm's-length from the political concerns of ministers. They are needed to ensure that water companies have sufficient finance for well-appraised investment, to facilitate competition, to achieve an efficient structure of prices.

Independent regulators have been able—under their statutory powers—to act both transparently and dispassionately. They do this within a framework of good process—a process able to involve all stakeholders in a situation where there are no simple answers.

Let us use the available tools to achieve a sensible balance by keeping water in the market economy.

REFERENCES

Byatt, Ian, 'The Water Regulation Regime in England and Wales', in C. Henry, Michel Matheu, and Alain Jeunemaître, *Regulation of Network Utilities: The European Experience* (Oxford: Oxford University Press, 2001).

DTI, *A Fair Deal for Consumers: Modernising the Framework for Utility Regulation* (London: Department of Trade and Industry, 1998).

OFWAT, *Paying for Growth* (Birmingham: Office of Water Services, 1993).

—— *Report on Conclusions from OFWAT's Enquiry into the Performance of Yorkshire Water Services Ltd.* (Birmingham: Office of Water Services, 1996).

—— *The Proposed Acquisition of Dwr Cymru Cyfyngedig by Glas Cymru Cyfyngedig* (Birmingham: Office of Water Services, 2001).

—— *Access Codes for Common Carriage* (Birmingham: Office of Water Services, 2002a).

—— *Linking Service Levels to Prices* (Birmingham: Office of Water Services, 2002b).

—— *Proposals for the Modification of the Conditions of Appointment of Anglian Water Services* (Birmingham: Office of Water Services, 2002c).

—— *Proposals for the Modification of the Conditions of Appointment of Southern Water Services Limited* (Birmingham: Office of Water Services, 2002d).

OFWAT, *The Proposed Takeover of Southern Water Services Ltd. by First Aqua Holdings Ltd.* (Birmingham: Office of Water Services, 2002*e*).

—— *Water Regulation: Facts and Figures* (Birmingham: Office of Water Services, 2003).

Treasury, Cmnd. 1337, *The Financial and Economic Obligations of the Nationalised Industries,* (London: HMSO, 1961).

—— Cmnd. 3437, *Nationalised Industries: A Review of Economic and Financial Objectives* (London: HMSO, 1967).

—— Cmnd. 7131, *The Nationalised Industries* (London: HMSO, 1978).

5

Water for Europe: The Creation of the European Water Framework Directive

Maria Kaika

WHEN the French politician Clemenceau visited Athens in 1899, he was taken on a tour of the city and briefed on the social, political, and economic problems facing both the city and the young Greek state. Afterwards, he addressed the local political and intellectual elites, starting his speech by exclaiming: 'The best politician amongst you shall be the one who will bring water into Athens' (Clemenceau 1899, cited in Gerontas and Skouzes 1963: III).

Indeed, water supply was one of the most important and intricate political and social issues of the nineteenth century. Although water supply and management is today often presented as a purely technological and engineering problem, it remains, as we shall see, a deeply political issue, implicated in relations of social power (Reisner 1990; Postel 1992). Indeed, today, more than a century onwards from Clemenceau's comment, his aphorism still holds true. Despite the fact that Western economies have undergone a period of 'fierce modernization' during the twentieth century, and despite technological advances and innovation, water supply and management remain major socio-technical issues at the heart of the political agenda (Bank 1992). Whilst contemporary Europe is not faced with severe water shortages (although many areas, particularly but not exclusively in the European South still face disruptions in water supply during dry months (ETC/IW 1996; ICWS 1996)), water supply and management remain amongst the most important political issues at the European and international level (Hundley 1992; Faure and Rubin 1993; Gleick 1993). Today, if anything, the political ecology of water has become more complex, and more important politically than in the nineteenth century. With the increasing internationalization and complexity of water resource management, with the emergence of an increasingly larger number of actors and institutions involved in this process, with the newly vested economic interests in water supply, and with the increasing concern and sensitivity towards environmental protection, if Clemenceau were alive today, he would probably maintain his aphorism— rephrasing it for the contemporary era: 'The best politician amongst you shall be the one who will bring clean water into Europe, while keeping happy all the parties involved in water supply, use, and management, at the local, regional, national, and European level.'

Such an objective may sound like an impossible task: however, it is a good description of what the European Commission undertook when it embarked on the creation of the European Water Framework Directive (WFD), a legally binding document that provides a common framework for water management and protection in Europe and promises to utterly transform the European water sector. The document was voted in by the European Union's Plenary Session in September 2000 and came into force in December 2000.

This chapter examines how the often conflicting interests of the new set of institutions, actors, and levels of governance that have replaced the traditional state-led approach to decision-making, along with the new economic, social, and environmental interests vested in water management affected decisions on the WFD. Rather than engaging with the scientific-technical innovations of the Directive, it examines the final text as the culmination of the social, political, and economic interests at the local, regional, national, and European levels that shaped and tailored the scientific debate and formulated the binding objectives of the WFD. In doing this, it argues that the question of who participated in the decision-making process and how they did this are central to understanding the final outcome.

A BRIEF HISTORY OF EUROPEAN WATER POLICY

The development of the European legislation for water resources can be grouped into three 'waves' (see Fig. 5.1). The first wave goes back to 1975 (Kallis and Nijkamp 2000) when the Surface Water Directive and the Drinking Water Directive were enacted. Those first directives focused predominantly on water quality standards and on the protection of surface waters that are allocated for abstraction (Da-Cunha 1989). The second wave of European water legislation came in 1991 and focused, for the first time, not only on setting acceptable water quality standards, but also on controlling emission levels as a means of achieving those desired standards. The new legislation included the Urban Waste-Water Management Directive, the new Drinking Water Quality Directive, the Nitrates Directive, and the Directive for Integrated Pollution and Prevention Control.

The WFD comprises what is now known as the 'third wave' of European water legislation and, in many ways it combines the two preceding

FIRST WAVE OF LEGISLATION
Focus on water quality objectives (WQO)

| 1975 | The surface water directive |
| 1980 | The drinking water directive |

SECOND WAVE OF LEGISLATION
Focus on emission limit value approach (ELV)

1991	Urban Waste-Water Management Directive
1991	Nitrates Directive
	New Drinking Water Directive
1996	Directive for Integrated Pollution and Prevention Control

THIRD WAVE OF LEGISLATION (THE WATER FRAMEWORK DIRECTIVE)
Integrated approach

February 1996	Commission Communication on European Water Policy
February 1997	Commission produces Proposal for a Water Framework Directive (COM(97) 49)
November 1997	Commission amends proposal following consultation (COM(97) 614)
January 1998	Commission involves environmental NGOs in amending Annex V on the proposed WFD
February 1998	Commission further amends proposal following consultation (COM(98) 76)
June 1998	Council of Ministers adopts a provisional common position on the WFD
Summer 1998	Environment Committee of the European Parliament amends the proposed Water Framework Directive and reveals substantial differences between Council of Ministers and European Parliament over the text of the WFD
Autumn/Winter1998	European Parliament deliberately procrastinates over giving the WFD a first reading in order to achieve co-decision status
January 1999	Informal conciliation talks under the auspices of the German Presidency of the EU between European Parliament, European Commission, and Council of Ministers
February 1999	European Parliament gives draft WFD its first reading—votes to accept 120 of the amendments made by the Environment Committee to the Commissions text
Summer 1999	Legislative process delayed by elections for European Parliament. European Commission accepts many of the amendments made by the European Parliament, but the Council of Ministers does not and reverts to the political agreement of June 1998
Autumn/Winter 1999	Environment Committee of the European Parliament re-tables its proposed amendments (PE 231.246) knowing the WFD will have co-decision status
February 2000	European Parliament gives draft WFD its second reading, accepting the bulk of the amendments proposed by the Environment Committee, challenges the common position adopted by Council of Ministers
May 2000	First round of formal conciliation talks between EU institutions unsuccessful
June 2000	Second round of formal conciliation talks produce a compromise Water Framework Directive
September 2000	The text drawn up in the conciliation talks was formally approved by a plenary session of Parliament and by the Council of Ministers
December 2000	WFD (Directive 2000/60/EC) published in the official gazette (22 December 2000, L 327/1), member states have three years from this date to transpose it into their legislation

FIG. 5.1 A chronology of the European legislation on water.

Source: Compiled by the author; comments by Dr Ben Page.

approaches. First, it introduces a new approach to water management based on river basins (an integrated approach), linking for the first time physical planning with water resource planning. Secondly, it stipulates that water quality cannot be seen outside emission controls and groundwater protection (a combined approach). Once it becomes fully operational, the WFD will replace all the water directives that are currently operational, namely:

1. The Urban Waste-Water Management Directive
2. The Nitrates Directive
3. The Dangerous Substances Directive
4. The Bathing Water Directive
5. The Surface Water for Drinking Water Abstraction Directive
6. The Directive on the Measurement of Surface (Drinking) Water
7. The Groundwater Directive
8. The Freshwater Fish Directive
9. The Shellfish Directive
10. The Drinking Water Quality Directive
11. The Information Exchange Decision
12. The proposed Integrated Pollution Prevention and Control Directive
13. The proposed Ecological Quality of Water Directive.(European Commission 2000*b*)

Development of the WFD began in 1995 when the Environmental Commission of the European Parliament (EP), the Council of Environment Ministers of the European Union (CM), and the Environment Commission (EC) agreed to embark upon a more global approach to water policy (WWF 2000). After this agreement, the Commission conducted a first draft communication (European Commission 1996) for a new water legislation. The original aims of this legislation, as they were stipulated in that first communication, were to:

1. Replace existing legislation with one comprehensive piece of legislation
2. Establish common definitions in EU water policy
3. Integrate water resource management
4. Integrate water quality and water quantity management
5. Integrate surface and groundwater management
6. Integrate measures (e.g. emission control) with environmental objectives (e.g. water quality).

RESPONDING TO CHANGES: THE MULTIPLICATION OF ACTORS, POWER CENTRES AND SCALES OF DECISION-MAKING AND THE NEED TO REFORM EU WATER POLICY

The decision radically to reform EU water legislation took place within a rapidly changing political, economic, and social framework. The principal actors and key social relations were also rapidly transforming, and this, in turn, made a new legislative framework necessary. We can identify three major parameters of change in the way that water is perceived, used, and managed within the last two decades.

The first is the multiplication of the actors involved in water management and the reconfiguration of their respective roles. Social power relations, conflicts, and debates over water supply are becoming increasingly intricate and complex. They are often vested with strong economic interests (Gottlieb 1988; Goubert 1989; Anon 1994) and have increasingly immediate social and political effects (Swyngedouw 1997). To start with, the growth of urban areas, the expansion of their ecological footprint, and the need to harness water from further away (often crossing national boundaries) has generated the need for regional and international agreements for water sharing and management. It has also generated a need for new institutions to manage such agreements. But, perhaps more importantly, the liberalization and subsequent internationalization of water markets has introduced the private sector as a new and powerful player in the field of water resource management and distribution. This same process of privatization has created the need for heavy institutional regulation (Neto 1998), thus generating an increasingly complex set of actors and institutions, such as governmental and industrial organizations, needed to regulate and control the water market (Frederiksen 1992; Saleth and Saleth 2000). In the case of the UK for example, the structure of water management has become significantly more complex after the asset privatization in 1989. Water supply projects are no longer one part of a state-led development of the collective means of consumption, they are also opportunities for market development, dealt with according to the 'laws' of the market economy and regulated through new institutional structures.

The second parameter of change is the multiplication of power centres and the scales at which decision-making is exercised in the water sector. This is an immediate effect of the multiplication of actors and of the changes in their respective roles (Ernst 1994; Swyngedouw, Kaïka, and Castro 2000). The complex system of institutions and actors, needed to deal with water management at the local, national, European, and international scale, relocated water politics, economics, and management from the sphere of the local into the sphere of the global (Ogden 1995; Swyngedouw 2000). This, in fact, reflects and complements a more general international reconfiguration and rescaling of power centres, the emergence of the European Union itself being one of them. This rescaling of decision-making is also part of the shift from a centralized, Keynesian, state-led, and state-controlled management (*government*) to a post-Keynesian management based on fragmented decision-making clusters (*governance*) (Jessop 1997). These clusters structure formal and informal relations, sometimes bypassing the national state in their decision-making. This does not mean that the relations among the different levels and clusters of governance are free of power formations. Indeed, the power configuration among these groups lies at the heart of their debates and is far from being static (Harvey 1989). For example, recently an important shift of power relations between the European Council of Ministers and the European Parliament took place when the Amsterdam Treaty was enacted. Later in the chapter, we shall examine how this affected the decision-making process for the WFD.

Finally, another important parameter is the increasing concern for the environment. Environmental protection, hardly a consideration at all in the first stages of industrial urbanization, now features centrally in debates about water supply and management at all levels of governance (ETC/IW 1998; ECEWTF 1997; European Commission, 1992; EEA 1995; 1998; 1999*a*, *b*, *c*). For example, today, new dam projects in European countries cannot be approved unless accompanied by an environmental impact assessment. A large amount of 'social capital' (Pretty and Ward 2001) has accumulated and is now invested in environmental protection and management. This comprises Non-Governmental Organizations (NGOs), quasi-Non-Governmental Organizations (quangos), institutions, and regulatory bodies, as well as civil groups and networks of people whose loyalties lie predominantly, if not solely, with the protection of the environment. Along with the emergence of these bodies, a new international elite of 'environmental experts' has emerged that is attached to NGOs, governmental bodies, and the private sector. The discourse and agendas of these environmental groups and organizations are in constant dialogue (opposition or accordance) with local, national, and international economic and political agendas.

The emergence of this new set of scales, actors, and relations has profound effects on decision-making procedures and on the ways in which political concord or opposition is voiced. In fact, a whole new way of 'doing politics' has emerged, whereby *political action* in its traditional form (i.e. protests, strikes, barricades, etc.) is giving way to practices of *participation* (Kearns 1995). We shall now see in detail how the above changes in actors, institutions, and social power relations filtered into the debate for the making of the WFD.

CHOOSING INTERLOCUTORS: POTENTIAL *v.* ABILITY TO PARTICIPATE

The tendency to substitute political action with participation is particularly strong in the decision-making process at the European level. This is partly to compensate for the difficulty of performing direct political action at the European level. In fact, the final text of the Water Framework Directive stipulates that there *must* be 'active public involvement' in river basin management planning. This, however, neither guarantees a fully inclusive participatory process, nor does it exclude the implication of relations of social power in the ability of each actor (or stakeholder) to participate. Although the European Union asserts its commitment to involve the public in the decision-making and implementation phases of its directives, practices of participation are not institutionally defined and neither are the roles of different political actors (e.g. professional organizations, NGOs., etc.). Thus, the question of who participates, where, and how, remains open and becomes a major source of political debate. Nevertheless, the configuration of the participating actors and their respective roles in the decision-making process affects the final decisions (e.g. directives) in an important way.

The obvious participating actors were the European Commission, the European Parliament (EP) and the European Council of Ministers (CM). Immediately after the first communication on the WFD, the European Commission launched an open call for participation at the drafting of the directive. This meant that, potentially, everybody could participate. However, in parallel to launching the open call, the Commission targeted a large number of specific groups and organizations and *invited* them to participate. They spanned: water suppliers; the chemical and fertilizer industry; the agricultural sector and farmers unions; specific NGOs, such as the European Environmental Bureau (EEB); regulators; and the water industry in countries with privatized water services. So, outside the open call for participation, as one of its members put it during an interview, the Commission 'chose its interlocutors' (interview, September 2000). Having said that, the selected interlocutors represented a wide range of conflicting interests, meaning that the Commission could not be accused of bias towards one particular political interest. It should also be noted that the Commission remained fully open to suggestions and inputs from every actor or organization throughout the consultation process, whether the actors in question were invited, or whether they participated out of their own initiative through such bodies as local authorities, landowners associations, statutory agencies, or consumers' associations.

For those groups who were not directly targeted by the Commission, the question of participation remained a question of dissemination of information. The EU has confirmed its commitment to dissemination of information on environmental issues through a Directive on Access to Environmental Information (90/313/EEC). This 'seeks to grant the public access to information on the environment which is held by public authorities or government controlled bodies with public responsibility for the environment' (European Commission, 1998: 25). However, despite the fact that the call for participation was open and publicly available (it was, for example, published on the European Union's web page) dissemination of information was still a problem, since one had to know the call was there in the first place, in order to follow it up. The groups who had the best access to information regarding the directive were the ones who had 'their man in Brussels'. Indeed, groups and organizations who had a Brussels bureau (such as: Eureau (European Union of National Associations of Water Suppliers and Waste Water Services), EEB (European Environmental Bureau), WWF (World Wildlife Fund), Greenpeace, ECPA (European Crop Protection Association), EFMA (European Fertilizers Manufacturers Association)) did much better than others in participating and raising their voice in the process of drafting the main objectives of the WFD.

However, the geographical location of those groups' headquarters is the outcome rather than the cause of their relatively more powerful position with respect to smaller groups and organizations. Indeed, Brussels-based groups are the ones who can afford the resources to have a Brussels bureau (and thus acquire direct access to lobbying and decision-making) and a dedicated person to follow the intricacies of the decision-making process on the WFD. Within the new 'governance' regime, access to lobbying gains even more political importance (Kearns 1995) and location becomes a significant factor for the successful promotion of political agendas. So, groups with enough resources to afford a Brussels bureau and a dedicated person were the ones who most benefited from the participatory mechanism of the European Union. Thus, well-funded and experienced groups had a 'structural' advantage in the participatory process, which was theoretically open to everybody. Practically, however, few groups and organizations could afford to follow and pursue the process all the way through. In what follows, we shall see how the dominant participating groups affected the decision-making process towards the WFD.

Since participation is a very complex concept that could mean anything from sharing decision-making to being informed of what has been decided (Pretty and Hine 1999), we need to clarify that *participation* in the making of the WFD meant *consultation*, i.e. participating groups could voice their opinion and consult the Commission (although the Commission was not obliged to take on board suggestions and views). They could not, however, share in the decision-making (Burkitt and Ashton 1996; Harrison *et al.* 2001). Participating actors' interests and concerns could be communicated to the Commission in writing or by voicing them in special conferences on the WFD organized by the Commission. Actors could also organize networks, discussion forums, and conferences, some of which were co-funded by the Commission. The final decision as to whether these suggestions would be incorporated or not lay with the Commission, but the more resources and effort groups put into voicing their suggestions and putting forward their agendas, the more likely they were to have their agendas incorporated into the proposal for the directive. As has previously been mentioned, the agendas that were expressed in this process represented a wide spectrum of opinion—from the fertilizer industry to environmental NGOs. In the next section, we shall identify the most important points of conflict between the various stakeholders. These were present from the very first stage and were contested until the very end.

MULTIPLE LEVELS OF CONFLICT AND INCOMPATIBLE AGENDAS

Although it is not easy to find common denominators for the wide variety of groups and interests represented, Fig. 5.2 is an attempt to classify the clusters of interests within the main groups of actors, with respect to three main controversial points of the directive: full cost pricing, hazardous substances and the implementation timeline.

Environmental NGOs were heavily involved in the process of drafting the WFD. NGOs did not always have the same agendas or priorities: the RSPB for example focused mainly on the impact on wetlands, while the EEB and the WWF pushed forward the agendas of the cessation of emission of hazardous substances and groundwater protection. However, environmental NGOs did have common points of interest for which they lobbied rather successfully, such as: the incorporation of the Esbjerg declaration (cessation of discharges of hazardous substances) (Meyer 1988); the incorporation of the OSPAR treaty (zero emission of priority hazardous substances); stricter implementation timelines; and a focus (controversially, at times) on full cost pricing as a means of environmental protection.

Strong lobbying points by actors involved							
	NGOs	Regulatory bodies	Local authorities	Consumers' organizations	Private water industry	Chemical industry	Agricultural sector
Hazardous substances	■	■				■	■
Full cost pricing	■			■	■		
Short implementation timeline	■		■			■	■

FIG. 5.2 Classification of clusters of interests among main actors with respect to the three main controversial points of the Water Framework Directive.

Source: Compiled by the author.

Quangos and Regulators under privatized water markets welcomed and strongly supported the WFD as a means of facilitating, complementing and enforcing regulation and environmental protection. Martin Griffiths, from the UK Environment Agency, noted that the WFD is: 'intellectually, exactly what we want'. River basin management in particular is considered to be a unique opportunity to develop further co-ordination of the actions of public authorities (Henton 2000). There is, however, concern amongst the regulators about the integration of the new regulatory regime imposed by the WFD with the existing regulatory bodies and institutions. Recently, the Environment Agency (UK) noted that:

Although the WFD represents an excellent opportunity for the protection and improvement to the water environment, because it is broad in scope and effectively overhauls the existing water management regime, its implementation will significantly affect the way in which the Agency carries out its business. Whilst the precise nature of the impacts are unclear, implementation will have consequences for the management of water quality, water resources, conservation, fisheries, flood defence, planning and environmental monitoring functions of the Agency.   (Wood 2001)

Local authorities had very mixed feelings about the directive. Although many local authorities responded to the initial call for consultation in the form of sending consultation papers to the Commission, they did not lobby during the conciliation phase. This was partly due to a lack of direct representation at the European level but also because they felt confident that their interests were adequately represented through their respective ministers. Many of them, as shown in interviews conducted (November 2000), simply were not aware of the WFD, or they felt it had nothing to do with them. However, as more and more local authorities have come to realize that the WFD will affect them considerably, they have adopted a more informed and active approach during the implementation phase. The directive, with its requirement for river basin management, through the establishment of River Basin Authorities, in effect recasts the relationship between physical, political, and administrative boundaries. For this reason it represents a thorny issue for local authorities in many European countries, which see statutory planning (including issuing abstraction licences) as one of their major mechanisms for exercising power. Thus, local authorities have deep concerns regarding the loss of power to the new administrative structure dictated by the WFD. Some go as far as to declare the 'death of statutory planning as we know it' (Gilbert 2000).

Consumers' organizations were not heavily involved in the proceedings of the directive. Although active in the first consultation phase, expressing mixed feelings about the directive (better water quality is considered highly desirable but concerns are expressed about the consumer having to take the financial burden for this and for environmental protection) they retreated from the process later on, partly due to the fact that they decided to put their resources into other campaigns (such as food labelling). However, they too call for a state-led solution to the subsidies necessary for environmental protection (Page 2000: 26).

The public *water supply sector* was mainly represented through local and national governments, but *the privatized water industry* pursued its interests both at the national level through lobbying environment ministers and, at the European level, through their European association, *Eureau*. Predictably enough, the industry supported full cost pricing, an issue on which they had the full support of NGOs. The prospect of better water quality, is particularly desirable for the water industry, since, amongst other benefits, it will reduce treatment costs. However, the industry has clearly pointed out that their thesis is that the financial burden for achieving better environmental standards should fall on the 'consumer', or on the state, and not on the industry itself. The environmental standards management of Anglian Water noted that 'The WFD will clearly drive further investment and the cost implications of this need to be transparent and fully understood. . . . Ultimately, the UK will be required to fund the Directive. . . . The polluter pays principal means, in this case, the water industries' customers' (Bolt 2000).

This is a very liberal interpretation (if not a subversion) of the EU's polluter pays principle, which does not contend that the cost should be taken directly by the consumer but stipulates instead that: 'one way to . . . avoid the occurrence of damage to the environment is indeed to impose liability on the party responsible for an activity that bears risks of causing such damage. This means that . . . the party in control of the activity (the operator), who is the actual polluter, has to pay the costs of repair.' (Article 174 (2) of the EC Treaty, see also European Commission 2000*a*). Perhaps representative of this attitude that consumers should be the ones to bear the burden of environmental protection, Water UK, the representative body of the UK water industry, responded to the UK government's first consultation on the implementation of the European Commission Water Framework Directive (2000/60/EC) by stipulating that:

[the Directive] could drive very significant new costs for water operators over the period 2005–2015. Indeed Annex B suggests that up to 40% of the total costs could fall on the water industry. Ofwat has a duty to finance the functions of companies, so these costs will be passed on to the industry's customers. It is therefore essential that we are fully involved in the implementation process so that the most cost effective options and solutions are examined and selected. (Water UK, 17 July 2001, 'Water Framework Directive: response to UK Government consultation')

Another concern for the water industry has been the issue of abstraction licences. Given that obtaining new abstraction licences is more economical for the industry than, for example, increasing leakage control, the industry is always keen on expanding its abstraction base. However, the directive's potentially strict standards for obtaining abstraction licences make such practices more difficult. Given the undisputed importance of reducing abstraction, the water industry in the UK recently replied cunningly to environmental considerations with other environmental considerations by marshalling climate change rather than over-abstraction as a possible cause for saline intrusion, thereby asking for climate change to be taken into consideration while setting abstraction limits:

We are concerned at the wording used in section 8.6 of the consultation. Saline intrusion is not only a symptom of 'overabstraction' but can also be due to sea level rise resulting from climate change. It is our view that the term 'overabstraction' should be used in the context of abstraction above the licensed allocation. . . . The use of the term 'overabstraction' may pre-judge the outcome of studies to evaluate the best option to deal with any issue.   (Water UK, 17 July 2001 'Water Framework Directive: response to UK Government consultation')

The chemical industry's main concern lies with the combined approach of the directive, namely the combination of environmental quality standards with emission limit values. The industry will be affected heavily by the directive's requirement for phasing out priority hazardous substances within twenty years. Thus, the chemical industry's association was actively involved in the final process of identifying priority hazardous substances pushing forward its own agenda for moderating the requirement for zero emission of such substances, in contrast to the NGOs' agenda for incorporating as many substances in the list as possible. Commenting on the directive's final text, the representative of the chemical industry's association contended at the WFD conference in London in December 2000 that 'the directive goes much further than expected' (Hackitt 2000). The same representative called for a 'realistic' [?] definition of zero [!] emissions of priority hazardous substances. A point with which the chemical industry finds itself in accord with the water industry and NGOs is full cost pricing, as it argues that environmental costs, including the cost for cessation of discharges of priority hazardous substances (which the industry estimates for the UK only between 200 and 1,000 million euros) should be met by the European citizen and not by the industry.

The agricultural sector is also potentially heavily affected by changes in water management (Garrod and Willis 1994). It would be wrong to treat the agricultural sector as one uniform agent, since it comprises very powerful and financially robust lobbies as well as thousands of small units who struggle financially (the latter category counts 70,000 units in the UK alone). Given, however, that the agricultural industry is the main diffusion polluter, there have been several issues unifying the sector's response to the WFD, the main being the concern about the directive's groundwater protection requirements and about potential increases in the price of water. The industry recognizes that diffusion pollution represents both a loss to the farmers and a waste of resources, but argues that it should be tackled through primary resource management (i.e. water and soil) and through educating farmers further about pesticide use (Tompkins 2000). As far as water resources are concerned, they consider themselves to be 'consumers'. Thus, they argue that the promotion of full cost pricing and other financial instruments for environmental protection find them at the end of the line further increasing their financial problems (they have put forward arguments about food security). However, they are in accord with both the chemical industry and the water industry that the financing of the changes required, if the WFD were to be fully implemented, should burden the state or the consumer (of agricultural products).

Apart from the above actors and their respective interests, *member states* within the European Union had varying national economic, political, and social interests which led to very different national positions with respect to the WFD. It should be noted, however, that what was expressed at the European level as 'national agendas' (i.e. what the ministers of member states promoted at the European Council) were compromises between contrasting interests of lobbies at the national level: for example, between the water and the agricultural industry over water pricing; between the chemical industry and the water industry over water quality; and between pressures at the local level for environmental protection versus pressures for further development of the agricultural sector. Although the WFD has been accused of being a 'Northern European Directive' because of its focus on water quality rather than water quantity issues, during the debates between the EP and the CM the traditional north/south divide was barely evident. For example, Ireland, which had introduced a new charging system with zero domestic charges only one month before the proposal was adopted, allied with the European south against full cost pricing. Portugal, on the other hand, allied with the European north on the issue of strict river basin management, since the imminent Spanish national hydrological plan will affect Portuguese water flows.[2] Such atypical alliances did not merely form at the nation state level: in what follows, we shall see how the conflicts and alliances between different member states and different lobbying actors and organizations culminated in the positions that the European Council and the European Parliament took in response to the Commission's proposal.

EUROPEAN PARLIAMENT *v.* EUROPEAN COUNCIL AND THE IMPORTANCE OF THE AMSTERDAM TREATY

May 1996 marked the end of the official consultation process and, in February 1997, the commission drafted the first proposal for a European Frame-

work Directive on Water (European Commission 1997). The main objectives of this proposal were to:

1. Expand the scope of water protection (surface and groundwater)
2. Achieve 'good status' for all waters (by a certain deadline)
3. Introduce an *integrated approach* (water management based on river basins)
4. Introduce a *combined approach* (emission limit values and diffusion pollution along with quality standards)
5. Get the prices 'right'
6. Get the citizen more closely involved
7. Streamline legislation.   (From Bloech 1999)

The proposal received very positive comments from all actors involved
and was particularly welcomed by environmental NGOs (Boymanns 1997).
The standard institutional procedure from that moment onwards was that
the CM and the EP had to conduct separate readings of the proposal and
suggest amendments that would then have to go back to the Commission
which would accept or reject them. After the Commission's assessment of
the amendments, the text had to go back to the EP and the EC for voting.
Although sounding like a mere bureaucratic exercise, this process involved
deeply complex negotiations. Soon after the Commission offered the proposal to the European Parliament and the Council of Ministers for reading
and amending, it became clear that the two decision-making bodies disagreed on a number of key issues. The EP, being more detached from, and
therefore more resilient to national networks of influence, supported far
stricter requirements for environmental protection and implementation
timelines. The CM, on the other hand, trying to juggle the often conflicting
interests of national industries, implementation costs, public organizations,
etc. adopted a much more lenient approach (see Fig. 5.3).

The three major points of conflict that remained controversial throughout the drafting process were:

1. The legally binding character of the directive's objectives (linked also to the implementation timeline)
2. The provision for cessation of release of hazardous substances (directly linked to the introduction of groundwater protection)
3. Water pricing (linked to full cost and environmental cost recovery).

Thus, the Parliament's position was almost antithetical to that of the
Council regarding the important issues of full cost recovery, hazardous
substances, and legislative binding objectives/timeline. At this point (1998),
however, it was the Council who had legislative power according to the
Maastricht Treaty, while the Parliament could only amend the Council's
proposed legislation after the Council had drafted it. The EP foresaw that if
the CM was to go ahead with voting its amended proposal, the WFD would

FIRST READING		SECOND READING	
COUNCIL (June 1998)	PARLIAMENT (February 1999)	COUNCIL (March 1999)	PARLIAMENT (February 1999)
Increase implementation period (16 years)	Keep implementation period. Oblige MS to report on progress	Increase implementation period (34 years). MS should 'make an effort' to implement WFD	Reduce implementation period (10 years)
Reject full cost recovery	Water is Europe's heritage, not a commercial product	Delete full cost pricing	Water is Europe's heritage, not a commercial product
Add list of derogations	No derogations	Increase list of derogations	Increase legally binding requirements
	Incorporate Esbjerg declaration. Identify priority hazardous substances for immediate cessation. Continuous reduction of all other hazardous substances	Abandon zero emission approach for hazardous substances	Fully incorporate OSPAR

FIG. 5.3 Comparative table of the positions of the European Parliament and the European Council of Ministers during the two readings of the Water Framework Directive.

present a much weaker environmental legislation than the set of directives it would repeal. In order to prevent this from happening, the Parliament made a very important political manœuvre and decided not to consider the WFD before the Amsterdam Treaty (AT) came into force on 1 May 1999. The AT (signed in 17 June 1997) radically altered the status of the decision-making procedure in the European Union from a cooperative process between the Council and the Parliament to a co-decision process between the Council and the Parliament[3] (see Figures 4 and 5).

By shifting the power balance between the EP and the CM, the Amsterdam Treaty gave equal negotiating powers to the Parliament, thus making the importance of the disparity of views between the EP and the CM even greater and thereby profoundly affecting the final text of the WFD and other similar environmental legislations (Bär and Kraemer 1998). An issue of political prestige was also involved in the EP's decision not to consider the directive before the AT became operational. According to the EP's rapporteur at the time, Ian White, the Members of the EP took offence at the fact that the EC reached an agreement on their stance before they had even listened to the Parliament's position.

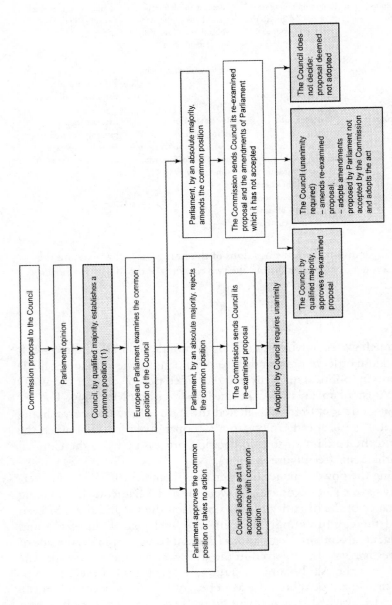

(1) The Council acts unanimously if it amends the proposal made by the Commission

Fig. 5.4 EU decision-making procedures: *Co-operation* (Article 252 EC Treaty).

Source: The European Commission.

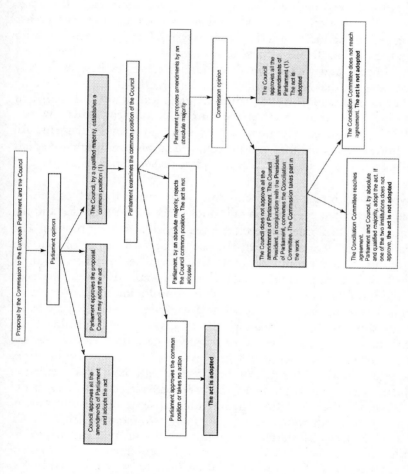

Fig. 5.5 EU decision-making procedures: *Co-decision* (Article 252 EC Treaty).

Source: The European Commission.

CONCILIATION TALKS: INTERNAL CONFLICT OR THE CULMINATION OF INCOMPATIBLE AGENDAS?

The disagreement between the EP and the CM, which may appear like an 'internal' conflict between the two main decision-making bodies of the European Union, was in fact the culmination of conflicting social, economic, and political interests involving the actors at different geographical and administrative scales described above. Indeed, although at this stage the official consultation process had finished, an unofficial but very intense lobbying period had just begun (Richardson 1997). Given the clear disparity between the EP and the CM, most actors involved tried to maximize their lobbying impact and to manage their human resources best by heavily lobbying one or the other (rather than both) official decision-making body. For example, industry (agricultural, chemical, water) lobbied their national ministers (thus, in effect the European Council) while environmental NGOs lobbied the members of the European Parliament. All parties, however, lobbied the Commission, which was attempting to redraft the directive and to respond to the amendments suggested by the EP and the CM.

Between the summer of 1998 and May 1999 (when the AT came into force) unprecedented conciliation talks took place between the EP and the CM on the WFD in order to accelerate the decision-making process. The conciliation talks were not a great success. In only three out of fourteen points of disagreement could a compromise be reached. Predictably, these were the least controversial ones: the inclusion of wetlands in the directive's scope; the introduction of rules for public consultation and marine conservation; and the inclusion of endocrine disrupting chemicals in the list of hazardous substances. The main points of disagreement remained: the implementation timeline; the end of release of hazardous substances; the promotion of full cost pricing as a strategy for environmental protection; and the legally binding character of the directive. Although the conciliation talks did not result in full agreement, they were considered to be 'very useful' by those who participated through providing a sense of the positions of different actors (Barth 2001). In February 1999 (two months before the AT came into force), the EP started elaborating on its own amendments to the Commission's proposal. The EP came up with 200 amendments (most of which were linguistic clarifications) and 133 of these were accepted by the Commission. The most important ones were the suggestions to:

1. Incorporate the Esbjerg declaration in the directive (rejected by the Commission)
2. Identify priority hazardous substances for immediate cessation (accepted)

3. Proceed with continuous reduction of all other hazardous substances (accepted)
4. Characterize water as Europe's heritage, and not a commercial product (rejected)
5. Oblige member states to report on the implementation progress (accepted).

It should be noted that, as with the Council's position which was an amalgamation of the positions of different member states and national lobbies, the Parliament's position was by no means a unanimously developed concurrence. Rather, it was an amalgamated majority consensus. For example, although the Parliament's agreed position was in favour of full cost pricing, the socialist MEPs of the south of Europe sided with the CM against full cost pricing, and characterized this clause as 'fundamentally neoliberal' because of the 'market economy' orientation of the argument and of the economic burden it would place on farmers of the European south and on consumers of water (i.e. all citizens) throughout Europe.

The Parliament's amendments did not receive too negative a response with the Commission (see the above list for the Commission's responses) but, when the amended text went back to the Council for a second reading, it faced major disagreements. The stakes were rising at this stage and lobbying from the different groups and organizations became fierce towards all three decision-making bodies: the Commission, the EP, and the CM (interviews with EEB, WWF, and Commission members, October 2000). In its second reading in March 1999, the Council amended the Parliament's positions by:

1. Rephrasing what used to be 'member states are obliged' to achieve good water status into 'member states should make an effort'
2. Deleting the requirement for full cost pricing
3. Increasing the implementation period (34 years)
4. Abandoning zero emission approach for priority hazardous substances
5. Increasing the list of derogations (exemptions from legally binding objectives).

With the above amendments in place, the WFD came very close to an 'empty word', with very few legally binding objectives incorporated into the text (European Commission, 1999). The Parliament, as well as the environmental NGOs, launched an immediate and strong reaction to the Council's amendments (interviews with members of the EEB, WWF, and of the Commission, October 2000; see also: European Parliament, 1999; European Environmental Bureau (EEB), Press Release, 12 March 1999).

At this point, there was a risk that the whole project for creating a new Water Directive for Europe would be abandoned altogether unless the EP and the CM could reach an agreement and come up with a commonly acceptable text. For this reason, the Environmental committee of the EP launched a conciliatory bid in January 2000 (just before they went ahead with their second reading) by accepting to drop their requirement for full cost pricing but stating that they were not prepared to make any further concessions. The bid was rejected by the CM and the EP went ahead with their second reading in February 2000 in which they voted for:

1. The full incorporation of the OSPAR Treaty into the WFD, which meant that all discharges of hazardous substances should stop by 2020
2. A 10-year implementation period
3. An increased number of legally binding requirements.

Given the differences between the EP's and the CM's positions, a second round of conciliatory talks became inevitable. At this point, the project for the directive was seriously at stake. According to the formal EU procedures, the two bodies had to reach an agreement within a deadline of six weeks or else the project would be dropped altogether. During this period, the Council, the Parliament, and the Commission engaged in heavily charged talks and lobbying, along with all the actors that had originally participated. Environmental NGOs in particular (notably the European Environmental Bureau (EEB) and the WWF) (Biliouri 1999) were lobbying and organizing meetings and discussions in an attempt to reach a political agreement without too many compromises on environmental protection.

THE GREAT SURPRISE AND THE GREAT COMPROMISE: THE COMMON TEXT FOR THE WATER FRAMEWORK DIRECTIVE

On 28 June 2000, after an exhausting conciliatory meeting behind closed doors between the EP and the CM, which went well into the early morning hours, an agreement on a common text was reached against all odds (European Commission 2000*c*). In order to produce the joint text, both sides made significant compromises and, although access to the minutes of the meeting is denied, we can read the compromising character of the talks in the adopted common text, which has now become the WFD. The final text's compromises on the main points of controversy are:

1. Member states should 'aim' to achieve good water status
2. The implementation deadline was to be set at 15 years
3. The controversial groundwater protection article was dropped and the requirement for a 'daughter directive' on groundwater protection was incorporated into the main text as an obligation

4. Priority hazardous substances should be eliminated 20 years after the publication of a list defining which substances fall into this category (European Parliament 2000; 2001*a*).
5. Full cost recovery should be 'taken into account' (European Parliament 2001*b*)
6. Member states can opt out of environmental cost recovery
7. Water was defined as Europe's heritage and not as a commercial product.

In September 2000, the directive's common text came into the Plenary Session of the European Parliament where it received final approval. The WFD came into force on 22 December 2000. From that date onwards, member states were given three years (until December 2003) to transpose the WFD into national legislation, facing strict financial fines if they failed to do so. However, as has been shown, since the final text is the product of a compromise, it is more of a tool than a strict piece of legislation (Kallis and Butler 2001; Lanz and Scheuer 2001). Most of the very controversial passages have been hedged in such a way that different member states can interpret them very differently. This makes the implementation phase immensely important, and gives particular executive powers to the actors who will participate in this phase.

THE IMPLEMENTATION PHASE: REPLACING CIVIL SOCIETY BY NGOS AND ADDRESSING THE SOCIAL ASPECTS OF ENVIRONMENTAL PROTECTION

Since there is no clear institutional structure to define participation at the implementation phase (European Commission 2001) participation and its perquisites are up for grabs. So far, the actors who are participating at this phase are more or less the same as the ones at the drafting and amending phases. Also, the channels for dissemination of information are more or less the same, resulting in a similar situation, whereby actors with more resources have a better chance of remaining informed. For example, the private sector (water industry, chemical industry) is much more successful in gaining access to information and is thus much more active at both the European and national levels. In contrast, the public water sector is far behind in keeping up with the changes and requirements. The same holds true for the majority of local authorities around Europe. At the member state level, different countries have demonstrated different degrees of interest in participating in the implementation strategy working groups. Spain, for example, has been one of the most active member states so far, not least because of its strong interest in the effects of the directive on the Spanish National Hydrological Plan.

Maria Kaika

It should be noted that in the implementation phase, as well as in the drafting phase there is a strong tendency to replace civil society with NGOs (environmental and others) who function increasingly as a token for public participation at the European level. This practice poses a number of problems.

First, there is a discrepancy between the general public's perception of what an NGO is (i.e. a non-governmental organization whose main aims are environmental protection) and of what constitutes an NGO in EU terminology (i.e. *any* non-governmental organization, regardless of its aims and scope). For example, when the Commission states that NGOs participated in a process, this can mean anything from water industry associations (qualifying as non-governmental organizations) to fertilizer industry organizations (equally non-governmental), to environmental NGOs. Secondly, by leaving public participation to NGOs, there is a risk that issues of social justice and equality (Seligman, Syme, and Gilchrist 1994) can easily fall through the net of cross-cutting interests of different organizations. The righteous and important focus from the part of environmental NGOs on environmental protection is often geared towards establishing technical and scientific principles for environmental protection (such as limit values, water quality standards, etc.) rather than establishing standards for social justice related to environmental issues (Harvey 1996). Having said that, it should be noted that large NGOs such as the EEB and WWF have significantly shifted their focus from technical issues to those of public interest and participation during the implementation phase of the WFD (Harrison *et al.* 2001*b*). Thus, in so far as there is no NGO directly addressing the particular issues of social justice linked to environmental protection, the debate around 'socially efficient sustainability' falls out of the agendas of a system of representation of civil society through NGOs.

Unless social issues are addressed alongside environmental issues, socially detrimental practices such as job losses in the water industry, significant increases in water prices, or selective network expansion and maintenance, can easily be justified in the name of 'the environment' or of 'sustainability'. As we have seen, already the chemical industry, the privatized water industry, and the agricultural industry demand that the 'consumer' (i.e. the citizen) bears the cost of environmental protection (Hassan 1995). Industry also actively promotes a populist discourse around social/political dilemmas between a cleaner environment versus 'social costs'. The social costs of environmental protection, according to the industry, may include higher prices for consumers of water, food, and chemical products (Hackitt 2000; Bolt 2000), 'sustainable food production', and promotion of redundancy policies to compensate for an increase in operational costs (Littlechild 1988; Maloney and Davidson 1994). However, a different set of 'social costs' related to environmental degradation (i.e. health hazards in the case of the chemical industry and saline intrusion from over-abstractions) does not feature as a parameter in the industry's cost benefit analysis.

RESURRECTING THE STATE: WHEN LAISSEZ-FAIRE BECOMES AIDEZ-FAIRE

As we have seen, the full range of industries pays lip-service to the importance and value of the WFD, asking the nation state to step in and provide a solution to their 'dilemmas' by taking on the financial implications of environmental protection stemming from the implementation of the WFD. The market's celebrated hand fails to reach and provide a solution for the environment, instead reaching only as far as its own profit lies, and being thrown up in despair when it comes to addressing the social and environmental cost of its acts. When it comes to paying part of its profit for its acts, the praised laissez-faire becomes aidez-faire and the exiled state is called in to pay the bill and solve the problems and dilemmas. For example, between 1989 and 1995 it was the British government who invested £17.9m. pounds to meet the new quality standards set by the EU's water directives. It was not the newly privatized water industry, which made billions of pounds' worth of profits during that same period, only a fraction of which was invested in environmental protection. However, under a regime of privatized assets, such as the UK's water sector, with a fragmentation of responsibilities and resources it is very difficult (although not impossible) for the state to earmark profits from a privatized industry for environmental protection and policies of maintenance and development of urban infrastructure (Cox and Jonas 1993; Clark and Root 1999).

In conclusion, as we have seen, there are a large number of thorny issues currently under discussion at the EU and at the member state level on the implementation of the WFD. It is clear, however, that while the dominant rhetoric promotes a neo-liberal agenda and a laissez-faire attitude in its declaration of the death of the state, at the same time, the state has become a magic agent called upon to solve the problems and to pay the costs of both industrial development and environmental protection. It is also, bizarrely, expected to take the burden of risk management while the industry can continue with business as usual. Water supply management and protection remain as thorny an issue as ever, and indeed, the title of the best politicians today *is* warranted to the ones who will bring clean water into Europe. However, in order for them to qualify, they also have to live up to the challenge of simultaneously making Europe's society more egalitarian through the reduction of poverty levels. In short, the social aspects of environmental protection must be addressed, without which it becomes meaningless. At the other end of the spectrum, the title of Europe's best citizen is warranted to those who get involved and do not leave their lives and the protection of their environment to be decided by others (be it NGOs, quangos, or local governments) by being mere spectators to the debates, or, worse, by being oblivious to political processes and decisions altogether.

ENDNOTES

1. The research for the paper that was the origin for this chapter was carried out during hte first phase of the EU Framework V project 'Achieving sustainable and innovative policies through participatory governance in a multi-level context', and was made possible thanks to funding from DG Research. Many thanks to Dr Erik Swyngedouw and Dr Ben Page, my colleagues at the School of Geography and the Environment, University of Oxford, who collaborated on this project and commented on the paper, and to Alex Loftus for his comments on earlier versions.

    The original paper 'The Water Framework Directive: a new directive for a changing social, political and economic European Framework' is published in *European Planning Studies* 2003 11(3): 299–316.

2. The plan includes the transfer of 100,000 m^3 of water a year from the basin of the River Ebro, in the north, to another four basins in the drier eastern part of the country. Of the water transferred, 55 per cent will be used for irrigation and 45 per cent for tourism and domestic use. The total cost of the plan is €23bn., a third of which is expected to be financed by the EU (WWF Global Network, 7 September 2001).

3. Before the Amsterdam Treaty came into force, and according to the Maastricht Treaty, the Parliament could either amend the Council's draft legislation (the so-called co-operation procedure) or withhold its 'assent' to Council decisions in certain areas (residence rights, the Structural and Cohesion Funds, Treaties of Accession, and others). The Amsterdam Treaty considerably increases Parliament's responsibilities by making the co-decision procedure more or less the general rule. The co-operation procedure will survive only within the confines of economic and monetary union. (EU publication: 'Effective institutions for an enlarged Europe'.)

REFERENCES

Anon., 'Free market for the land of the free?', *Water and Environment International*, 3 (1994), 14–15.

Bank, W., *World Development Report 1992: Development and the Environment* (Oxford: Oxford University Press 1992).

Bär, S. and Kraemer., A., 'European Environmental Policy after Amsterdam', *Journal of Environmental Law*, 10 (1998), 315–30.

Barth, F., 'Working with the Water Framework Directive at the European Level', presentation to workshop *The Freshwater Framework*, Globe EU FIMENEL 19 September 2001

Biliouri, D., 'Environmental NGOs in Brussels: how powerful are their lobbying activities?', *Environmental Politics*, 8 (1999), 173–85.

Bloech, H., 'The European Union Water Framework Directive: taking European water policy into the next millennium?', *Water, Science and Technology*, 40 (1999), 67–71.

Bolt, S. A. W. 'Water Supply and Disposal' in *Water Framework Directive Conference*, The Barbican, London, 7 November 2000: UK CEED (Centre for Economic and Environmental Development). Water UK.

Boymanns, D., 'The EEBs position on the proposal for a Water Framework Directive' (Brussels: European Environment Agency, 1997).

Burkitt, B. and Ashton, F., 'The birth of the stakeholder society', *Critical Social Policy*, 16 (1996), 3–16.

Clark, G. L. and Root, A., 'Infrastructure shortfall in the United Kingdom: the private finance initiative and government policy', *Political Geography*, 18 (1999), 341–65.

Cox, K. R. and Jonas, A. E. G., 'Urban development, collective consumption and the politics of metropolitan fragmentation', *Political Geography*, 12 (1993), 8–37.

Da-Cunha, L. V., 'Water resources situation and management in the EEC', *Hydrogeologie*, 2 (1989), 57–69.

Ernst, J., *Whose Utility? The Social Impact of Public Utility Privatization and Regulation in Britain* (Buckingham, Phil.: Open University Press, 1994).

European Commission, *Towards Sustainability: A European Community Programme of Policy and Action in Relation to the Environment and Sustainable Development*, COM(92) 23, Final Volume II (Brussels: European Commission, 1992).

—— *Proposal for a European Parliament and Council Decision on an Action Programme for Integrated Groundwater Protection and Management*, COM(96), 315 Final (Brussels: European Commission, 1996).

—— *Proposal for a Council Directive Establishing a Framework for Community Action in the Field of Water Policy*, COM(97), 49 Final (Brussels: European Commission, 1997).

—— *Guide to the Approximation of the European Union Environmental Legislation*, revised from SEC(97) 1608 of 25 August 1997 (Brussels European Commission, 1998).

—— '*Proposal for a Council Directive Establishing a Framework for Community Action in the Field of Water Policy*, COM(99), 271, 17 June 1999 (Brussels: European Commission, 1999).

—— *White Paper on Environmental Liability*, 9 February 2000, COM (2000), 66, Final (Brussels European Commission, 2000*a*).

—— *Handbook on the Implementation of EC Environmental Legislation* (COM(97), 49 Final, 30 October 2000) (Brussels European Commission, 2000*b*).

—— 'Directive 2000/60/EC of the European Parliament and of the Council of 23rd October 2000 establishing a framework for Community action in the field of water policy', *Official Journal*, 22 December 2000, L 327/1 (2000*c*).

—— *Common Strategy on the Implementation of the Water Framework Directive*, DG Environment, Strategic Document, May 2001.

European Commission Environment Water Task Force (ECEWTF), Working document presented at the validation workshop on water research priorities for Europe, Baveno, Italy, 1997.

European Environment Agency (EEA), *Europe's Environment—The Dobris Assessment*, ed. David Stanners and Philippe Bourdeau (Copenhagen: EEA, 1995).

—— *Europe's Environment: The Second Assessment* (Copenhagen: EEA, 1998).

—— *Groundwater Quality and Quantity in Europe,* Environmental Assessment Report, 3 (Copenhagen: EEA, 1999*a*).

—— *Environment in the European Union at the Turn of the Century.* (Copenhagen: EEA, 1999*b*), ch. 3s. 5.

—— *Sustainable Water Use in Europe—Sectoral Use of Water,* Environmental Assessment Report, 1 (Copenhagen: EEA, 1999*c*).

European Environment Agency (EEA) / European Topic Centre on Inland Waters (ETC/IW), *Water Resources Problems in Southern Europe,* European Environment Agency, Topic Report, 15 (1996).

European Parliament, 'Draft recommendation for second reading on the common position adopted by the Council with a view to the adoption of a European Parliament and Council directive on establishing a framework for Community action in the field of water policy', PE 231.246, 26 November 1999, http://www.europarl.eu.int/meetdocs/committees/envi/20000125/381796_en.doc, accessed 10 October 2003.

—— 'Draft report on the proposal for a European Parliament and Council decision establishing the list of priority substances in the field of water policy', PE 293.659, 8 August, 2000, http://www.europarl.eu.int/meetdocs/committees/envi/20001009/414353_en.doc, accessed 10 October 2003.

—— Legislative resolution on the amended proposal for a European Parliament and Council decision establishing the list of priority substances in the field of water policy (COM(2000) 47—COM(2001)), (2001*a*).

—— 'Draft report on the Commission Communication to the Council, the European Parliament and the Economic and Social Committee on pricing policies for enhancing the sustainability of water resources', PE 304.663, 31 May 2001*b*, http://www.europarl.eu.int/meetdocs/committees/envi/20010618/438985en.pdf, accessed 10 October 2003.

European Topic Centre on Inland Waters (ETC/IW) *Groundwater Quality and Quantity in Europe COM(96) 500, 22 October 1996* (report by A. Scheidleder *et al.*, Austrian Working Group on Water) EEA Monograph PO26/97/1, 1998.

Faure, G. and Rubin, J. Z., *Culture and Negotiation: The Resolution of Water Disputes* (London: Sage, 1993).

Frederiksen, H., *Water Resources Institutions: Some Principles and Practices,* (Washington, DC: World Bank, 1992).

Garrod, G., and Willis, K. G., *The Transferability of Environmental Benefits: A Review of Recent Research in Water Resources Management,* Centre for Rural Economy, Department of Agricultural Economics and Food Marketing, University of Newcastle upon Tyne, England, 1994.

Gerontas, D., and Skouzes, D., *The Chronicle of Watering Athens,* (Athens: n. pub. 1963).

Gilbert, C. K. C. C., 'Integrating the Directive with the Statutory Planning Framework', in *Water Framework Directive Conference,* The Barbican, London, 7 November 2000: UK CEED.

Gleick, P.H., *Water in Crisis: A Guide to the World's Fresh Water Resource*. (Oxford: Oxford University Press, 1993).

Gottlieb, R., *A Life of Its Own: The Politics and Power of Water* (San Diego: Harcourt Brace Jovanovich, 1988).

Goubert, J. P., *The Conquest of Water* (Cambridge: Polity 1989).

Hackitt, J. C. I. A., 'Hazardous Substances', in *Water Framework Directive Conference*, The Barbican, London, 7 November 2000: UK CEED.

Harrison, A. *et al.* 'WWF's preliminary comments on public participation in the context of the Water Framework Directive and Integrated River Basin Management', WWF European Freshwater Programme, Copenhagen, 2001.

Harvey, D., 'From managerialism to entrepreneurialism: the transformation in urban governance in late capitalism', *Geografiska annaler*, 71 (1989), 3–17.

——*Justice, Nature and the Geography of Difference* (Oxford: Blackwell, 1996).

Hassan, J., 'The impact of EU environmental policy on water industry reform', *European Environment*, 5 (1995), 45–51.

Henton, P. S., 'Implementing the Directive: the view of the regulator', in *Water Framework Directive Conference*. The Barbican, London, 7 November 2000: UK CEED.

Hundley, N., *The Great Thirst* (Berkeley and Los Angeles: University of California Press, 1992).

International Centre of Water Studies (ICWS), *Long-Range Study on Water Supply and Demand in Europe—Integrated Report*, ICWS, Amsterdam, Netherlands. Report 96.05 to the EC-Forward Studies Unit, 1996.

Jessop, B., 'Capitalism and its future: remarks on regulation, government and governance', *Review of International Political Economy*, 4 (1997), 561–82.

Kallis, G., and Butler, D., 'The EU water framework directive: measures and implications', *Water Policy*, 3 (2001), 125–42.

Kallis, G, and Nijkamp, P., 'Evolution of EU water policy: a critical assessment and a hopeful perspective', *Journal of Environmental Law and Policy*, 3 (2000), 301–55.

Kearns, A. 'Active Citizenship and Local Governance—Political and Geographical Dimensions', *Political Geography*, 14 (1995), 155–75.

Lanz, K., and Scheuer, S., *EEB Handbook on EU Water Policy under the Water Framework Directive* (Brussels: EEB, 2001).

Littlechild, S., 'Economic regulation of privatized water authorities and some further reflections', *Oxford Review of Economic Policy*, 4 (1988), 2.

Maloney, W. A., and Davidson, J. O., 'Privatization and employment relations—the case of the water industry—Davidson, Jd', *Public Administration*, 72 (1994), 615.

Meyer, U., 'The EU Water Framework Directive: The Importance of Incorporating Esbjerg Declaration', *Oceanographic Literature Review*, 45 (1988), 1428.

Neto, F., 'Water privatization and regulation in England and France: a tale of two models', *Natural Resources Forum*, 22 (1998), 107–17.

Ogden, S. G., 'Transforming frameworks of accountability—the case of water privatization', *Accounting Organizations and Society*, 20 (1995), 193–218.

Page, B., *Results of the Case Study on the Water Supply in London and the UK. Project Report: Achieving Sustainable and Innovative Policies through Participatory Governance in a Multi-Level Context* (Oxford: School of Geography and the Environment, 2000).

Postel, S., *The Last Oasis: Facing Water Scarcity* (London: Earthscan, 1992).

Pretty, J., and Hine, R., 'Participatory appraisal for Community assessment: principles and methods', Colchester: Centre for Environment and Society, University of Essex, 1999).

Pretty, J., and Ward, H., 'Social capital and the environment', *World Development*, 29 (2001), 209–27.

Reisner, M., *Cadillac desert: the American West and its Disappearing Water* (London: Penguin, 1990).

Richardson, J., 'EU water policy: Uncertain agendas, shifting networks and complex coalitions', *Environmental Politics*, 3 (1997), 139–67.

Saleth, R. M. and D. A., 'Institutional changes in global water sector: trends, patterns, and implications', *Water Policy*, 2 (2000), 175–99.

Seligman, C., Syme, G. J., and Gilchrist, R., 'The role of values and ethical principles in judgments of environmental dilemmas', *Journal of Social Issues,* 50 (1994), 105–19.

Swyngedouw, E., 'Power, nature, and the city. The conquest of water and the political ecology of urbanization in Guayaquil, Ecuador: 1880–1990', *Environment and Planning A,* 29 (1997), 311–32.

—— 'Authoritarian governance, power and the politics of rescaling', *Environment and Planning D: Society and Space,* 18 (2000), 63–76.

——Kaïka, M., and Castro, E., 'Urban Water: A political-ecology perspective', *Built Environment* 28/2 (2002), 124–37

Tompkins, J. (National Farmers Union Policy Adviser), 'Diffuse Pollution' in *Water Framework Directive Conference,* The Barbican, London, 7 November 2000: UK CEED.

Water UK, *Water Framework Directive: Response to UK Government Consultation,* 17 July 2001.

Wood, A., 'Preparing the Agency for the implementation of the Water Framework Directive', Environment Agency UK, 2001.

WWF, *EU Water Framework Directive: A History of Negotiations and Lobbying 1999–2000, last update on 13 November 2000,* WWF, Living waters programme, 2000. www.panda.org, accessed 18 January 2004.

—— 'Analysis and economic valuation of the Ebro river transfers in the Spanish National Hydrological-Plan', WWF Global Network Mediterannean Programme, 25 July 2003: http://www.panda.org/news_facts/publications accessed 18 November 2003.

6

Water Laws for Water Security in the Twenty-First Century

Stefano Burchi

INTRODUCTION

Concern for the long-term sustainability of water resources development and use has gained definitive prominence on the agenda of the world community at the Second World Water Forum and Ministerial Conference held at The Hague in March 2000. The concept and goal of *water security* were loosely articulated there, by reference to 'key challenges', namely, meeting basic needs; securing the food supply; protecting ecosystems; sharing water resources; managing risks; valuing water; and governing water wisely. *Governance*, in particular, attracted attention and debate at the International Freshwater Conference held in Bonn, in December 2001, preparatory to the United Nations World Summit for Sustainable Development (Johannesburg, 2002) and to the Third World Water Forum (Kyoto, 2003). Governance has also attracted the attention of the water ministers of African countries meeting in Abuja, Nigeria, in April 2002, and it has been echoed in the ensuing *Abuja Ministerial Declaration on Water* committing African countries to put in place 'arrangements for the governance of water affairs at all levels'.

It is readily apparent that water security, and the governance issues which that concept and goal trigger in train, will be the mainstay of much contemporary international and domestic discourse about water resources. However, the authoritative pronouncements recalled earlier invariably fail to pin down with accuracy the concepts of 'water security', and of 'good governance' in relation to water. The 'Recommendations for Action' issued from the Bonn Freshwater Conference articulate seventeen *priority actions* in the field of water-related governance at the domestic and the international levels, which, in effect, help substantiate that elusive concept and translate it into measurable goals. Implicitly, action, in particular, at the domestic level is underpinned by legislation for the management and development of water resources, setting out a web of rights and obligations for the resource users, for government, and the members of civil society.

This chapter will focus on such legislation, and, in particular, on the requirements for a supportive legal framework for the 'priority actions' recommended by the Bonn conference. A comparative state-of-the-art review and analysis of the contemporary legal framework for the management of water resources will first be made, and salient features and main trends highlighted. Next, assessment parameters will be drawn from the 'Recommendations for Action' issued from the Bonn International Conference on Freshwater. The contemporary legal framework for the management of water resources at the domestic level will be assessed against these parameters, and pointers offered helping to delineate the likely evolution of water resources legislation, in support of the concept and goals of water security and water-related governance.

SALIENT FEATURES AND TRENDS OF CONTEMPORARY DOMESTIC WATER RESOURCES LEGISLATION

Contemporary water resources management legislation is preoccupied with the fair and transparent allocation of available water resources for abstraction and use, and as a vehicle for the disposal of waste, with an eye to efficiency and equity of allocation and to ecosystem preservation. In particular, a comparative analysis of contemporary legislation discloses a number of discrete trends pointing in the above-mentioned direction.

Incorporation of water resources into the public domain of the state

Private groundwaters, and riparian rights in surface watercourses and in groundwater, have been steadily attracted into the ever-expanding sphere of 'public' waters. Public waters may include those waters held in the public domain (ownership) of the state, as in the water laws adopted in Italy in 1994, in Morocco in 1995, and in Zimbabwe in 1998, all of which bring all groundwater resources within the public domain. Alternatively, water resources have been vested in the state in trust for the public, such as in South Africa's 1998 National Water Act, or the state may be vested with superior user rights, as in Uganda's 1995 Water Resources Act and in the Australian state of New South Wales's Water Management Act of 2000. Whatever the legal underpinning, the result has been to bring all or most of the nation's water resources under the scope of the government's allocative authority.

Licensing by Government of Water Abstraction and Use

As a result of water becoming public property, individuals can only claim and obtain user rights in water. Such rights generally accrue from a grant made by the government and recorded in a permit, licence, concession, or like instrument. In some western states of the United States, it is the courts that make such grants, in adjudication-type proceedings. Grants tend to be time-bound, and are qualified by terms and conditions as to, notably, volume and rate of water abstraction, point of abstraction, return of excess flows, and payment of charges, which are entered in the permit, licence, or concession. Once made, grants are subject to review and adjustment to reflect new circumstances, including, notably, the need to accommodate a new user. In this instance, however, the grantee is entitled to receive compensation for the prejudice s/he suffers. Permits, licences, and concessions are subject to suspension and cancellation on account of, notably, non-use of the water and non-observance by the grantee of the terms and conditions of the grant, or of the provisions of the law in general. As a general rule and trend, the regulation of abstraction licensing seeks to reconcile the security of water rights tenure implicit in the government grant of an abstraction licence, permit, or concession, and the flexibility which is desirable to adjust allocation patterns to the ever-changing circumstances of water availability, of evolving abstraction and use technologies, and of shifting development and conservation policies and priorities, and political agendas.

Checking the government authority to allocate and reallocate water resources, thereby improving the quality of allocation decisions

The discretionary authority the government enjoys in making grants and allocating water has traditionally been checked by the courts of law or through the hierarchical review (appeal) opportunities available through the granting process. These traditional review mechanisms are available *after* an allocation decision had been made and a permit granted. By contrast, a new generation of fetters tends to operate *before* such decisions and the relevant grants are made.

These are basically aimed at improving the quality of decision-making, thereby obviating the need for review and litigation. The allocative authority of government is increasingly qualified by legislation imposing Environmental Impact Assessment (EIA) requirements in respect of proposed water abstractions; by water resources planning determinations, especially where these have a binding effect on government decisions; by the imposition of minimum flow requirements in respect of surface watercourses, which seek to protect the ecology and fish life of watercourses and act as a limit on the government's allocative authority by barring abstractions above established limits; and by the formal reservation of water quantities or flows for a specific purpose—notably, the satisfaction of basic human needs and the protection of ecosystems, which also puts limits on the government's allocative authority in that the water reserve cannot be allocated except for the reserved purposes.

Controlled trading in water rights

Increasingly, lawmakers have been turning to trading in water rights in the pursuit of efficiency of water allocation and use. Because trading of water rights empowers users to make allocative decisions instead of government, part of the expanding allocative authority vested in the government, which was observed earlier, is returned to the users.

Unregulated water trading is known to operate perhaps only in Chile. There, water is regarded as a commodity which can be freely traded through the sale of the relevant government grants. Elsewhere, regulations on water trading seek to minimize the possibilities of unwelcome 'third-party' effects, such as effects on the environment, on the interests of the area from where water is taken for use in another area, on cultural values, on resource availability to meet priority requirements and, generally, on marginal groups. There are countries, however, such as Morocco, where water has traditionally attached to the irrigated land where it is used, and where the trading of water rights separate from land rights is explicitly forbidden by the current (1995) Water Act.

Charging for water abstraction, and regarding water as an economic good

Charging for water abstraction (also known as 'user pays' principle), seeks to influence the demand for water and constitutes the chief non-regulatory mechanism available to control water abstraction and use. It is generally practised in combination with the regulatory mechanisms described earlier. The criteria governing rate-setting vary, from the relative scarcity of water resources and the different kinds of use, as in Mexico, to the volume, kind of use, location and source, as in France, to the recovery of the operating costs of the government water resources administration, as in England and Wales.

Curbing water pollution, in particular from 'non-point' sources

Well-tested regulatory and economic instruments for the prevention and abatement of water pollution from point sources (notably, industrial outfalls and municipal sewers) feature in many recent water pollution control laws. Such instruments range from discharge permits linked to effluent quality standards and quality objectives/standards for the receiving water source, to charging for discharging waste in water bodies (the well-known 'polluter pays' principle).

The more recent statutes bear evidence of a growing concern for pollution of, in particular, groundwater resources from diffuse sources, such as the run-off and drainage of cropland in rural areas. The focus of regulation has shifted from the discharge itself to the land use giving rise to a diffuse discharge. Thus, cultivation practices have been increasingly attracting regulatory restrictions aimed at preventing, abating, or minimizing pollution from substances such as the nitrates employed in agriculture. In December 1991, for example, the European Union adopted Directive 91/676 directing member states to designate nitrate-sensitive (or nitrate-vulnerable) areas and to draw up a code or codes of 'good agricultural practice'. Within the designated areas, the provisions of such code or codes become mandatory for farmers. Furthermore, the polluter pays principle mentioned earlier is making inroads into the agricultural sector, which has traditionally been impermeable to charging policies. In France, for instance, in 2001 legislation has been tabled in Parliament *inter alia* introducing a tax on nitrogen-based pollution of water resources, whether surface or underground, from agriculture.

Other legislation focuses not on the land use but on the area where the diffuse pollution is occurring and on specific activities that can cause or worsen pollution. In Spain, for example, under the current consolidated Water Act (2001), the government has the authority to declare an area experiencing groundwater pollution or the risk of it as a 'protected aquifer area'. In such areas, groundwater withdrawals may be limited or frozen pending the adoption of a recovery plan for the aquifer. A similar approach is reflected in the Water Management Act of 2000 of New South Wales, which mandates the formation by the local community of legally binding 'aquifer management plans'.

In France, the Water Act of 1992 imposes permit requirements on all wells and groundwater extractions in areas designated by the *préfet* as 'chronic groundwater shortage' areas, regardless of the amounts of water actually extracted. In the majority of states in the western United States, where groundwater extraction and use are governed by prior appropriation, 'control areas' can be established where applications for new groundwater extraction permits are no longer granted as a matter of course, but may be approved only after surviving a string of tests, hearings, and reviews.

Participation of water users in the management of water resources

The formation of groupings of water users for the development and management of sources of irrigation water is widely practised and regulated in most of Latin America, in Spain and Italy in Europe, and in many South Asian countries. Customary practices also play a dominant role in some jurisdictions, such as the island of Bali (Indonesia), and in many oases in the Saharan and Sahelian regions of Africa. Elsewhere, water users are called upon by the legislation to shoulder increasing responsibilities in the management of water resources under stress, and to make up the membership of the internal structure of the government water administration.

The direct involvement of users in the management of water resources under stress is a regular feature in much recent water legislation. Often, this is in connection with groundwater resources in areas experiencing accelerated depletion and/or severe pollution. In Texas, (US), Groundwater Conservation Districts are formed at governments' instigation in designated 'critical areas', i.e. areas experiencing overdraft, insufficient supply, or contamination. Whereas these districts have varied powers including issuing permits, spacing wells, and setting the amount of withdrawals, most have not imposed such regulatory mechanisms, instead opting for voluntary self-restraint and educational programmes. As such, they do not interfere with the landowners' rights to pump groundwater. Under New South Wales's Water Management Act of 2000, water users must be represented in the groundwater management committees established for the management of aquifers under stress.

In Spain, the 2001 consolidated Water Act provides for the compulsory formation of water users' groups from among the users of an aquifer, when the aquifer is being, or is at risk of becoming, overexploited. These groups are to share groundwater management responsibilities with the government, in particular in the management and policing of groundwater extraction rights. In Mexico since 1995 a number of Groundwater Technical Committees (COTAS) have been established under the auspices of the federal Water Authority, to allow the participation of users, together with federal, state, and local agencies, in the formulation and implementation of programmes and regulations for aquifer preservation and recovery. Water User Groups (WUGs) for the management of water supply points, and Water User Associations grouping any number of WUGs, are regulated by Uganda's 1995 Water Statute. South Africa's 1998 National Water Act also provides for the formation of water user associations from among those wishing to undertake water-related activities for their mutual benefit.

Water users and their interests are represented in the organs that make up the internal structure of river basin authorities and agencies. Thus, for instance, Spain's River Basin Authorities (*Confederaciones Hidrográficas*) include users' representatives in their decision-making and advisory organs. Similarly, users' representatives make up at least two-thirds of the total membership of the board of directors of France's Water Agencies (*Agences de l'eau*). They are also represented on the Agencies' advisory Basin Committees.

Irrigators hold a minority of seats on the board of directors of Morocco's new Basin Authorities, which are being formed pursuant to the 1995 Water Act. In South Africa, water users and environmental interest groups will be represented in the decision-making structure of the Catchment Management Agencies established under the new Water Act. A similar approach is reflected in Zimbabwe's 1998 Water Act, which provides for the establishment of Catchment and Sub-catchment Councils and for the representation of water users thereon. Under Brazil's 1997 federal Water Act, water users will be represented in the Basin Committees, alongside the representatives of civil society and of the federal, state, and municipal governments concerned. The functions of such committees are akin to those of their Mexican counterparts.

Interface between statutory and customary water rights

Customary law in many countries still plays an important role in water management, particularly at the community level. Customary water laws are rarely a single and unified body of norms, and vary widely from region to region, sometimes even between villages in the same region. Customary rules governing access to water have been documented in many countries, the best-known example being perhaps the allocation system of irrigation water and relevant water rights practised since time immemorial on the island of Bali, Indonesia. Another example of customary law is riparianism, which is or has been practised in a variety of forms in many common law countries.

Despite the social and economic significance of customary systems and practices, their interface with statutory law has seldom been mapped out and seldom is it reflected in the legislation. The Water Act (1974) and subsequent Irrigation Regulations (1982) of Indonesia are a rare exception, as they openly acknowledge the traditional system of irrigation water allocation and rights practised in the island of Bali, and grant it equal dignity to the statutory allocation and rights system inaugurated by the legislation. In a similar vein, the 1998 Irrigation Law of Bolivia recognizes as *de facto* organizations water users' groups formed on the basis of customary norms. Namibia is considering new water resources legislation to replace the 1964 Water Act in force. The new legislation being mooted seeks to map out the scope of the interaction between statutory water rights and customary rights in general, and to minimize opportunities for conflict. Under the new legislation being mooted, the government will be required to take due account of existing customary practices and rights in granting statutory water use permits. The documented existence of such rights and practices will not only influence the grant or denial of a statutory right; it will also attract special terms and conditions to be entered in the grant of a statutory right, for the specific purpose of protecting existing customary rights and practices.

ASSESSMENT

The 'priority actions' recommended by the International Conference on Freshwater held in Bonn, December 2001, are sufficiently precise that useful parameters can be drawn for a meaningful evaluation of contemporary water resources legislation against the water security and good water-related governance concepts and goals implicitly canvassed by the same Bonn conference recommendations. Whereas the Bonn 'actions' refer only occasionally to legislation, they imply the availability of a supportive and enabling domestic, legal/regulatory framework, providing in particular:

1. stable and transparent rules that enable all water users to gain equitable access to, and make use of, water;
2. water resources planning mechanisms;
3. mechanisms for the allocation of water resources that 'should balance competing demands and take into account the social, economic and environmental values of water', while ensuring efficiency of water use;
4. allocation mechanisms ensuring minimum flows through ecosystems 'at levels that maintain their integrity';
5. arrangements to 'protect ecosystems and preserve or restore the ecological integrity' of groundwater, rivers, and lakes;
6. 'effective legal frameworks' for protecting water quality (this is one instance where the Bonn 'actions' make explicit reference to legislation);
7. the participation of the 'people' in general, and of 'local stakeholders' in particular, in managing local water needs and resources.

With due allowance for the inevitable degree of generalization implicit in the review and analysis of salient features and trends worldwide, given in the section 'Salient Features and Trends of Contemporary Domestic Water Resources Legislation' above, it is readily apparent none the less that much contemporary water resources legislation, by and large, satisfies the requirements for an enabling and supportive domestic legal/regulatory framework. In particular, underpinned by the *incorporation of water resources in the public domain of the state* [italics denote subheadings of previous section], the current practice of, and legislation on, *licensing by government of water abstraction and use* seek to regulate access to water resources by all prospective users, and to provide a fair and transparent chance to all. Equity of access is an implied goal in the detailed regulation of the process of obtaining, generally from government, stable and secure rights in the resource. Efficiency of allocation and use, however, is equally prominent in much contemporary legislation regulating access and, in particular, the *licensing by government of water abstraction and use*. This concern is also apparent in the contemporary legislation providing opportunities for *controlled trading in water rights*, and in the legislation enabling *charging for water abstraction*, in recognition of the economic value of water. As a result, contemporary legislation regulating water resources abstraction and use licensing, trading in water rights, and charging for water abstraction and use, by and large meets the standards implicitly laid down in this respect by the Bonn recommended actions.

Water resources planning is a standard feature of contemporary water resources legislation, which moreover, tends to mandate the formation of plans at the basin level and to make planning determinations mandatory on government licensing of water resources abstraction and use. As a result, on this count too, contemporary legislation meets the Bonn actions standards. The prominence accorded in the Bonn 'actions' to water quality and to the need to preserve and restore it is amply reflected in much contemporary legislation *curbing water pollution, in particular from 'non-point' sources* and, to this end, regulating wastewater discharging and land uses—notably, cultivation practices. Also, the concern of the Bonn 'actions' for the protection of the environment and, in particular, of water-dependent ecosystems is well reflected in the legislation imposing Environmental Impact Assessment (EIA) requirements in respect of proposed water abstractions; in the legislation imposing minimum flow requirements in respect of surface watercourses, seeking to protect the ecology and fish life of watercourses and barring abstractions above established limits; and in the legislation enabling the reservation of water quantities or flows for ecosystem protection purposes, which puts the reserved water beyond the scope of the government's allocative authority.

Finally, the *participation of water users in the management of water resources* constitutes a well-entrenched feature of much contemporary water resources legislation on the formation of associations of water users, and on the composition and membership structure of the government water administration at the river/lake basin level. This is entirely in line with the last of the standards implicit in the Bonn actions.

CONCLUSIONS AND POINTERS FOR THE FUTURE

The agenda that has been emerging as water security and water-related governance goals have come into sharper focus, particularly at the United Nations World Summit for Sustainable Development (Johannesburg, 2002) and at the Third World Water Forum (Kyoto, 2003), will no doubt require supportive legal and regulatory frameworks at the domestic level. By and large, the contemporary legal and regulatory frameworks for water resources available domestically are supportive of the actions recommended at the Bonn International Conference on Freshwater, which offer the best approximation so far of water security and governance concepts and goals. Obviously, this does not mean that all is well and that there is no room for improvement and no scope for further evolution. Quite the contrary. On the one hand, water resources management legislation, however robust, serves no purpose if it is not implemented and vigorously enforced. Implementation and enforcement are areas of regulatory legislation in general most neglected and most in need of attention. However, both reach past the domain of the law alone, and beg issues of national policy and priority-setting, which are larger than the law. Secondly, demands for a supportive legal and regulatory framework are intensifying as water security and water-related governance goals and issues have come into sharper focus, at the global events recalled earlier. The debate about the future shape and direction of water resources legislation will, in all likelihood, delve into a few large issues, outlined below which emerge from the contemporary trends highlighted earlier in this chapter.

Reconciling security of tenure with risk and uncertainty

Water allocation mechanisms and relevant legal instruments are growing in complexity and sophistication, in an effort to reconcile the security of water rights tenure with changing socio-economic development policies and hydrological, hydrogeological, technological, and political circumstances— in other words, in the effort to force risk and uncertainty onto the stability of tenure and rights sought by water-sector investors. This will call for the creative use of classic regulatory instruments, which must be time-bound and adapted to rapidly changing circumstances with minimum conflict. This will require doing away with grants and rights of indefinite duration, fine-tuning grants to actual or predictable water availability, scaling grants downwards to reflect actual or probable user requirements and release the surpluses for further allocation by the grantee or by government. There is scope for fine-tuning in all these respects in much contemporary water legislation.

Pursuing opportunities for efficiency gains without neglecting equity

Where cultural and religious barriers do not oppose it, the use of market mechanisms opens up opportunities for efficiency gains in the allocation and use of water resources. This will require relaxing the rigidities built in water allocation systems based on government water abstraction licensing, where grants are tied to a specific use or location. Markets for water rights, however, must be regulated to protect third parties interests and such intangibles as environmental, amenity, and cultural values. Moreover, equity cannot entirely be sacrificed to economic efficiency. To a large extent, these concerns are reflected in a vast majority of the countries that have accepted the market as the proper mechanism for the allocation of water resources to competing uses. Giving equity its due, however, and adapting market mechanisms for the allocation of water resources to other cultural contexts, will be very challenging tasks ahead.

Empowering users to shoulder greater responsibilities

With a view to minimizing conflict, critical decisions for the adaptation of water allocation patterns and of water rights to changing circumstances—notably, with respect to resources at high risk of contamination or depletion—will need to be taken and monitored by, or with the participation of, the users concerned. This will require the availability of legislation enabling the formation of legally constituted groups, and empowering them through the devolution or the delegation of the required authority. Existing legislation is quite advanced on the former aspect, while delegation and devolution offer room for further development.

Granting the environment a high profile in allocating water resources to competing uses

In a fair and transparent process for the allocation of water resources among competing sectors and users through the available regulatory and economic instruments, the environmental requirements of freshwater bodies will need to be granted a standing and dignity equal to, if not higher than, development needs. To a large extent this is already happening through a number of regulatory instruments and statutory requirements that prioritize the health of surface and groundwater systems in the process of allocating water resources to competing uses, and qualify and restrict the authority of government, in its capacity as custodian of water resources and grantor of user rights in them.

Enhancing the government's and user's absorptive capactity

The emerging regulatory and non-regulatory allocation and pollution control mechanisms reviewed in this chapter tend to be so complex that they require a pretty sophisticated government water resources administration to operate them and to translate them into micro-level actual decisions, and a responsive water users population to collaborate in the successful operation of such mechanisms on the ground. This points to the obvious need to enhance the absorptive capacity of the government water resources administrations, and to the less obvious need to inform and to educate the water users populations, and to phase in the inauguration of sophisticated regulatory and non-regulatory and machinery in progressive fashion, as the capacity of both government and water users to absorb them increases progressively.

Mapping out the interface between customary and statutory water allocation systems

So far, customary water allocation systems have attracted, at best, the benign neglect of much contemporary water resources legislation. Yet, it stands to reason that opportunities for conflict between customary water allocation and use systems and allocation systems regulated by statute should be minimized, and that the full scope of the interface between the two systems should be mapped out if coexistence is to be pursued as a serious option. This is an area of water legislation that invites creative thinking, in the search for viable and workable mechanisms for the accommodation of traditional systems and for their coexistence alongside statutory systems of water resources allocation and use.

CONCLUDING SUMMARY

Public waters, i.e. water resources held by the public under a variety of legal configurations, have grown in scope to the point where one may legitimately wonder whether the laws and the reality on the ground bear out any longer the concept of private water—and whether the vexed issue of public versus private waters has any real meaning any more. The scope of governments' authority in the allocation and reallocation of public water resources to members of the public, under the instrument of the grant of usufructuary rights of enjoyment and use, has expanded in parallel. This expansion, and the flexibility needed by governments to react to shifting circumstances and priorities, are countervailed by checks and balances designed to restrain governments' authority and, eventually, to improve the quality of governmental decision-making. Significantly, some of the allocative authority vesting in the governments as a result of the ever-expanding scope of public water resources is being returned to the water users inasmuch as they are allowed to trade their water rights in response to the laws of the market. Trades, however, tend not to be unfettered, for governments are careful to retain varying degrees of involvement in the water users' market-driven allocation decisions.

Concern for the pollution of water resources, notably groundwater, from such diffuse sources as the run-off and drainage of intensively cultivated fields is being responded to through land-use control instruments. The regulator's attention is thus shifted from the discharge end of the spectrum to the land-based practices giving rise to the diffuse discharge.

Water users and their groupings and associations are increasingly being called upon to take management responsibilities at field level, notably in connection with irrigation water, and to take part in governments' water-related decision-making, through their membership in the internal structures of river basin authorities and agencies.

What lies ahead on the agenda of water laws for the new century is the further refinement of water allocation mechanisms, which must strike a dynamic balance between equity and efficiency in allocation and use. Water allocation structures and policies must reflect the uncertainties of water availability under regulated and unregulated flow conditions, while taking into account the security and dependability of water rights sought by users and investors. A further challenge is to reconcile the development of water resources with conservation and with protection of the quality of water bodies, not just for further use but also for the survival of water-dependent habitats. Users will have to be empowered to take up a higher profile in management decisions, especially under critical circumstances. Customary water law must come out of the legal limbo where it has been confined by, or as a result of, statutory water law, and contribute to the achievement of governance goals. Lastly, successfully meeting the challenges ahead will hinge heavily on the government water administrations attaining increasing levels of sophistication, and on the collaboration of responsive water user populations with government.

ENDNOTE

The author bears sole responsibility for the views and opinions which have been expressed in this document. Nothing contained in it commits or purports to commit the Food and Agriculture Organization of the United Nations.

REFERENCES

Burchi, S., '2001 year-end Review of comparative international water law developments', *Water Law*, 12 (2001), 330–7.

—— 'Current developments in water legislation', *Water Law*, 11 (2000), 110–16.

Caponera, D., *Principles of Water Law and Administration—National and International* (Rotterdam: Balkema, 1992).

FAO, *Water Rights Administration—Experience, Issues and Guidelines*, by H. Garduño Velasco, Legislative Study, 70 (Rome: FAO, 2001).

—— *Issues in Water Law Reform*, Legislative Study, 67 (Rome: FAO, 1999).

—— *Preparing National Regulations for Water Resources Management—Principles and Practice*, by S. Burchi, Legislative Study, 52 (Rome: FAO, 1994).

7

Water and Conflicts, Hobbes *v.* Ibn Khaldun: The Real Clash of Civilizations?

Julie Trottier

THIS chapter will begin by considering the two opposing schools of thought concerning water wars. A first school of thought has maintained since the 1980s that competition over water will lead to wars as relative water scarcity increases around the planet. A second school of thought has emerged as a response, arguing that competition for water, far from leading states to wage war on each other, will rather incite them to cooperate. The arguments of each of these schools of thought and the common hypotheses that underlie both sets of theories will be explored. The evolution of war in an era of globalization and of a state's involvement in competition for water will be examined, which will lead to revisiting the concepts of water wars and water cooperation. How the various theories of war that emerged from the three great Western ideologies, conservatism, liberalism, and radicalism, limited the definition of issues and the choice of factors that were deemed relevant when examining water conflicts will be studied. This chapter details how a Hobbesian prism was used to look on a Khaldunian reality, which has prevented us from understanding the coming water conflicts and has left us ill equipped to deal with them.

WATER WARS

'Water conflicts will cause the wars of the twenty-first century.' This is more than a catchy statement: it is the object of numerous arguments and counter-arguments in the scientific community as much effort has been devoted to either proving or disproving the causal connection between water scarcity and water wars.

Thomas Naff and Ruth Matson (1984: 181) seem to have launched the debate by arguing that 'water runs both on and under the surface of politics in the Middle East', and analysing the role played by water in riparian state relations. A series of publications followed that supported the concept of the causal link between water and war (Starr 1988, 1991) (Bulloch and Darwish 1993; Biswas 1994; Soffer 1999). The development of this literature led Hussein Amery (2001: 51) to refer to 'the well-established and thoroughly documented positive link between resource scarcity and violent conflict'. Clearly the idea of a causal link between water scarcity and war has grown over the past twenty years to the point where it could become ideologically hegemonic. Even Kofi Annan (2001) was declaring 'and if we are not careful, future wars are going to be about water and not about oil'. This illustrates that the concept was not confined to academic circles but was structuring the thoughts of high-level political officers. The idea according to which competition for water in water-scarce areas constitutes the greatest danger of war was growing to be taken as a given, an unquestionable fact of life.

This first school of thought has led to what Ohlsson (1999*b*) has called 'the numbers game'. As the causal link between water scarcity and war remained unchallenged, the relevant question appeared to be quantitative: how much renewable water existed within the boundaries of every state? How much constituted scarcity? Engineers and hydrogeologists such as Sharif El Musa (1996) produced numerous studies detailing the various quantities of water available to every state in arid zones, especially in the Middle East. M. Falkenmark (1989) pioneered the idea of a water stress threshold. The ratio of the quantity of renewable water within a state's territory over its population was held to be an indicator of water scarcity. Water security was achieved if the state contained more than 10,000m³ per capita. Water availability was deemed adequate if the state contained from 1,666 to 10,000m³ per capita. States endowed with 1,000 to 1,666m³ per capita were deemed to be water stressed. They were said to be chronically water stressed if they contained between 500 and 1,000m³ per capita, and to lie beyond the water barrier if they contained less than 500. This indicator of water stress was essentially based on an estimate of the quantity needed in agricultural production using irrigation. A state that could not be self-sufficient in food production was deemed to be water stressed although these per capita water quantities were sufficient to cover domestic needs.

Disturbing charts were drawn up, showing the various renewable water endowments of Middle East states (Beshorner 1992). According to such an indicator, Turkey, Lebanon, and Iraq were deemed to have adequate water supplies while Israel, Jordan, the West Bank, and the Gaza Strip lay beyond the water barrier. Such inequality was deemed highly dangerous as it was thought it could propel the water-poor states to wage war on the water-rich states. This became the topic of detailed study in international relations circles (Lowi 1993*a*, *b*) and social scientists followed suit by concentrating on how international law could contribute to a 'just' and sustainable water-sharing among states, suggesting various allocations among riparian states. (Benvenisti and Gvirtzman 1993). It is worth noting that the majority of the water-war literature focused on the Middle East.

WATER PEACE

A second school of thought emerged during the 1990s, denying the causality between water scarcity and international war. J. A. Allan (1998) developed the concept of 'virtual water' to describe the water necessary to produce imported food. Importing one ton of cereal was virtually equivalent to importing the corresponding quantity of water necessary to produce this cereal. Allan demonstrated that more 'virtual water' already flowed in the Middle East than real water flowed in the Nile. Indeed, by 1999 Jordan was already importing 91 per cent and Israel 87 per cent of their cereals (Postel 1999: 130). Food security does not necessarily entail food self-sufficiency, he argues. Calculating water-stress indicators on the basis of the agricultural production capacity does not allow us to predict the likelihood of war among states. Arid states have far more to gain from cooperation in keeping the price of cereals low than in wars to appropriate each other's water (Allan 1992).

In what is probably the most ambitious survey of water crises and treaties around the world carried out so far, Aaron Wolf (1998) argued that water has brought about much more interstate cooperation than conflict. He analysed 412 crises among riparian states between 1918 and 1994 and identified only seven cases where water issues contributed to the dispute (Wolf 1999). Empirical evidence thus seems to corroborate Allan's proposition.

Much of the water-war literature concentrated on the Middle East, especially on the Arab–Israeli conflict, and so did much of the water peace literature. Arnon Medzini (1997), focused on the link between water resources and the determination of the limits of the state of Israel. He argued that water did not play a role either in demarcating the mandate's border in 1923 or in determining the 1948 armistice line. Gershon Baskin (1994) calculated that, were Israel to buy in 1993 a quantity of water equivalent to that lying in the West Bank's aquifers, it would spend 0.67 per cent of its GDP. No state in its right mind would ever go to war for a stake that was worth so little, he said. The authors promoting this second school of thought argued that states facing water scarcity cooperate in order to solve their problems, simply because that is the most rational thing to do (Beaumont 1994, 1997). UNESCO launched a PCCP programme in 2000, 'from Potential Conflict to Cooperation Potential', in the hope of reversing the growth of the first school of thought and of persuading educators, decision-makers, politicians, and diplomats that water generated cooperation much more frequently than it did war.

SHARED ASSUMPTIONS

Both schools shared a set of assumptions that are worth examining. First of all, both approaches portrayed states as rational actors that decide whether or not to go to war on the basis of a rational analysis. Second, sovereignty over water was portrayed as a national interest. This implied that a state's national interest would be impaired if it lost sovereignty over water resources and furthered if it gained such sovereignty. Such assumptions have been challenged in a general manner by the constructivist school of thought that has emerged within the field of international relations theory since 1995 (Wendt 1995; Katzenstein 1996; Finnemore 1996; Barnett 1998). Constructivism argues that national interests and preferences are socially constructed. This makes them liable to change rather than set in stone. Ideas are important forces that shape preferences according to constructivists. Clive Lipchin's (2003) exploration of the role played by Zionism in Israel's perception of water as a national interest confirms this hypothesis. Moreover, constructivists insist on the fact that rationality is always contextual. Yet such comments do not seem to have been taken on board within studies on water and politics.

A third assumption seems to pervade the water-war literature. States are portrayed as controlling and managing water within their territories. States are assumed to be spelling out the rules governing water use, access, and allocation. Indeed, gaining or losing sovereignty over water resources is systematically portrayed as a loss or a gain in determining such use, access and allocation. Yet this is far from corresponding to reality in most of the developing world where many of the arid and semi-arid states lie.

Finally, a fourth assumption concerns the shape that conflicts over water might take: wars involving states as protagonists. Homer-Dixon (1991) has put forward an alternative model that allows us to consider other actors apart from states, but his contribution remained largely theoretical and the body of literature on water conflict did not integrate his model into its empirical studies. This chapter will now examine the shape taken by violent conflicts in our modern era of globalization. It will lead us to re-examine critically these four assumptions shared by the two schools of thought concerning water wars.

WARS IN AN ERA OF GLOBALIZATION

Wars are waged by states. Other forms of social organization can enter into violent conflict but, legally speaking, only states can wage war. A violent conflict pitting a state against a mafia, a state against a transnational network, a state against any other form of social organization can be described as a police operation, terrorism, genocide, or many other names. But it cannot be described as a war.

The emergence of the state as the dominant form of social organization took place in the Middle Ages and Renaissance period. The Treaty of Westphalia in 1648 offers a milestone to mark the advent of the state as the only type of actor recognized in international relations. This treaty ended the Thirty Years War and recognized the territorial state as the legal cornerstone of the modern international system. Joel Migdal (1988) argues that the state was able to supersede all other forms of social organization in medieval Europe because it was the most efficient at mobilizing the resources within a territory. This allowed the state to appropriate a monopoly over the exercise of legitimate violence. The state became the sole institution able to spell out the rules governing social control. This has led us to perceive the realm of international relations as featuring states as the only actors. At best, intergovernmental organizations were eventually considered as relevant actors by virtue of the fact that they were composed of states. Within such a cognitive map, violent conflicts in international relations could only involve states. The existence of other actors could not be perceived except as that of illegitimate bandits.

This cognitive map remained more or less applicable to international relations in the Western world from 1648 to 2001. In the last few decades, however, we have entered a period Anthony Giddens (1992: 149) calls radicalized modernity, where the consequences of modernity have simply become more radicalized and more universalized than before. Giddens defines modernity as a form of social organization or social way of life that emerged in Europe in the beginning of the seventeenth century and later acquired a global influence. He identified three dynamics driving this phenomenon. First of all, the separation of time and space allowed social actors to co-ordinate themselves in time and thereby control space. Second, the disembedding of social relations from their local contexts and their restructuration through indefinite time–space spans were carried out via two fundamental mechanisms. The appearance of symbolic tokens such as money, for example, allowed transactions to be lifted out from the specific exchange context where barter would have taken place. The use of expert systems led humans to trust technologies and forms of organization without understanding their functioning. Finally, the ordering and reordering of social relations via the reflexive appropriation of knowledge allowed modernity to come into existence.

Such transformations in our social organization were fuelled by technological innovations. Giddens points to the crucial role played by the advent of the mechanical clock in the separation of time and space. Initially, these changes were most felt in the realm of economic production. The greater mobility of capital and goods, and the instantaneous speed of communications around the planet led the economic actors to change their form of organization. Globalization describes the process whereby production and trade operate over the entire planet. It does not mean that consumption patterns become universal. For example, a greater market for tchadris in Afghanistan after the Taliban took power could spur entrepreneurs to have tchadris made in China, dyed in India, and sold in Afghanistan while their accounting would be carried out in Sri Lanka and they would be living on a yacht in the West Indies or in a townhouse in London.

Forms of social organization are forever evolving. The new types of links that emerged from globalization were soon put to use within several forms of social organizations. Grandmothers started emailing their grandchildren who lived half a world away while virtual scientific groups emerged, gathering researchers of many nationalities. People became used to interacting for a variety of purposes over extended lengths of time without ever meeting physically. Freedom fighters set up their websites and reported daily the abuses committed by their oppressors. Indeed over the last fifteen years, many forms of social organization as diverse as the family, research communities, and liberation organizations appropriated the methods previously used by economic actors within globalization.

A commercial organization, no matter how much it claims a specific 'enterprise culture', differs fundamentally from other forms of social organization. Indeed, the members of a commercial enterprise belong to it essentially for material gain. The entire commercial enterprise also operates essentially for material gains. Other motivations such as prestige in the case of a manager, or respecting the environment in the case of a company, are subordinated to the first material motivation: earning a salary in the case of the employee, generating profits in the case of a company. Other forms of social organization, in turn, often rely on other types of motivation. It can be love in the case of a family, or an ideal in the case of a research community or a liberation organization. Just as commercial enterprises had previously used mobility of goods and capital and instantaneous communications to further their aims, other forms of social organization also started to use the tools of globalization, and networks emerged.

Networks now prove to be a very efficient form of social organization to mobilize resources in order to achieve violence, on a par with the state. September 11, 2001 may very well become the milestone marking the end of the era when the state was the dominant form of social organization just as 1648 is the milestone marking the beginning of that period. Some Western states have recognized this and are now engaged in a ruthless struggle for survival with this new form of organization. In 2001, both the UK and the USA adopted draconian legislation that severely curtails long-established rights in an attempt to prevent network forms of social organization from mobilizing means of violence. It may be that this authoritative evolution will continue until the state finally expires as a dominant form of social organization.

Future violent conflicts are likely to pit networks against states and networks against each other. The era of interstate wars is drawing to an end by virtue of the fact that states are no longer the most efficient structure to mobilize violence. Why has the water-war literature focused only on wars, i.e. on violent conflicts among states? Much of this literature has been produced by engineers and hard scientists, who were not challenging paradigms defining the polities they were discussing. Western political science, in turn, has been deeply influenced by Hobbes's ideas concerning the state. This has led many social scientists to perceive other forms of social organization as vestiges of the past, doomed to disappear. This is the underlying thinking behind, for example, Fukuyama's *The End of History?* (1989)

Postmodernist criticism of social science finally allowed a break with this unilinear vision of history as it questioned the very prism through which scientists defined their research questions or made their observations. Unfortunately, this criticism often led to cultural relativism, concluding that it was useless to carry out research because knowledge was always contextualized.

This chapter examines how some Western ideas concerning water wars and states' interactions with water have been shaped by pervasive beliefs prevalent in our culture for centuries. Some of these beliefs may very well predate Hobbes and most of those sharing these beliefs have never read Hobbes, yet they are very clearly expressed within Hobbes's writings. His book, *Leviathan*, was first published in 1651 and was reprinted several times before its printing was discouraged. Hobbes influenced intellectuals of his time long before he came to be used by Bentham in the nineteenth century and by others later as a basis for modern utilitarian theories (Macpherson 1962). In fact, some of the ideas we find in Hobbes have pervaded modern Western thought far beyond the intellectuals' realm. Some now shape the paradigms within which we analyse water management and conflicts. These ideas have become hegemonic beliefs in the Gramscian sense of the term, and it is worth examining them in order to reassess the issue of water and conflicts.

HOBBES

Thomas Hobbes admired Euclidean geometry and built his political science using axioms in the same manner as geometry arguments are developed. His first axiom was that man is rational. His second axiom was that man wants first and foremost his self-preservation. All of Hobbes's theoretical construction then lies on these two axioms. He concluded that men will rationally give up their power to a third party on condition that everyone else do the same. Hobbes was writing at a time when medieval communitarian ties and forms of solidarity had long decayed and had failed to provide men with security. His main concern was how to achieve peace, a burning question as he was writing during the English Civil War. He argued that the old forms of communitarian ties had to be replaced by only one type of allegiance, that of the citizen to the state. This was the rational thing to do as only this could ensure peace among men who had a natural propensity to inflict violence on each other.

Hobbes deplored what is nowadays referred to as segmentary nationalism: the fact that an individual's allegiance goes first to his family, second to his clan, third to his tribe, and only last to his nation. 'For though they obtain a Victory by their unanimous endeavour against a forraign enemy; yet afterwards, when either they have no common enemy, or he that by one part is held for an enemy, is by another part held for a friend, they must needs by the difference of their interests dissolve, and fall again into a Warre amongst themselves' (1985: 225). Communitarian ties certainly were not portrayed as conducive to cooperation. Their potential for constructing social capital, i.e. the capacity a community has of acting collectively in order to accomplish a goal, was recognized only within warfare. Nowhere does Hobbes acknowledge communitarian ties as building blocks that are useful for successful social organization.

History seemed to prove Hobbes right at the time when he was writing. The Treaty of Westphalia was signed only three years before the publication of *Leviathan*. Indeed, the state emerged in Europe as the dominant form of social organization after the old forms of medieval solidarity had failed to provide security for its members. Hobbes's theory described this phenomenon quite adequately.

Such ideas have deeply shaped our thinking to this day. Rationality and mathematical methods are still regarded as the best way to analyse society. Quantitative methodologies are widely regarded as more reliable than qualitative methodologies. This contributes to explaining why so much of the water-war literature is devoted to quantitative assessments of water resources. Qualitative research would have allowed a challenge to the underlying assumptions that were used. But such research was not perceived as necessary or valuable by the scientific community itself.

More importantly such ideas about the necessary decay of communitarian forms of social organization, in fact of any type of social tie other than the state–citizen relation, have been at the root of developmentalist and state–periphery theories. Kinship ties, tribal ties, communal property regimes were all supposed to be 'traditional' and were expected to be doomed as we were meant to become modern. These social ties were based on emotion, religion, or family allegiances. The state–citizen tie was the only rational one, and therefore the only one appropriate to the modern citizen. This belief pervades the theory of the tragedy of the commons as well (Hardin 1968). Hardin portrays development as a unidirectional progression. He perceives only three types of property regime: open access, which he erroneously calls the commons, private property regimes, and public property regimes. He perceives only two possibilities to temper the use of natural resources: a switch from open access to either private or public property regimes. 'First we abandoned the commons in food gathering, enclosing farm land and restricting pastures and hunting and fishing areas. These restrictions are still not complete throughout the world,' writes Hardin (1968: 1248). Such a linear vision of history fails to perceive either the past or present role of communitarian forms of social organization in resource management via communitarian property regimes. This belief became so deeply ingrained that it prevented recognition of other forms of social organization that persisted in our own Western states where only the state–citizen tie was supposed to prevail.

Hard scientists and engineers are very sensitive to arguments based on rationality. They provided most of the water-war and water-peace literature without questioning such Hobbesian ideas. They also provided most of the water-policy literature, all the while hostage to the idea that the state was the sole actor that could spell out the rules concerning how water is managed.

WATER AND THE STATE

Hobbes rightly perceived the emergence of the state as a sole holder of the legitimate means of violence in Europe. Whether the state, as a form of social organization, achieved the same success in securing the sole exercise of social control in other domains remains highly questionable. European states certainly never achieved such a monopoly over the right to spell out rules concerning water management. Thierry Ruf (forthcoming) detailed the intricate interactions among the various communitarian irrigation institutions in the Pyrénées Orientales and between them and the French state over the last 700 years. Springs and wells all around the Mediterranean have often been controlled by communitarian institutions according to customary law. In the Netherlands and in Belgium, water user associations named *waterschappen* or *wateringue* must be consulted by the central administration for any proposal aiming to modify the existing situation. In Germany, the water protection associations have always played a role within the Lander, and the Water Tribunal in Valencia, Spain, has been functioning for over a thousand years. Customary law also remains very important in the Po Valley and in Sicily (Caponera 2000: 80).

In the developing world, Hobbes's ideas rarely match reality. The states that emerged from decolonization did not succeed in breaking down the existing social ties and in replacing them by a state–citizen relation (Badie 1992). They rather struck the best compromises possible with other forms of social organization. They did not achieve a monopoly over the right to spell out rules. Other forms of social organization persisted and continued to exercise social control (Migdal 2001). This social control was exerted to regulate issues as varied as trade, education, marriage, and, more importantly for the purposes of this chapter, water management (Trottier 1999). Another great thinker could have been used to describe this situation more adequately. Ibn Khaldun remains little known in Western society. His ideas certainly never pervaded Western thought as Hobbes's did. Considering them is worthwhile in order to re-examine our assumptions concerning war, water management, and water wars. His fame in the Arab world certainly rises to that of Hobbes in the Western world and he probably had an influence on the evolution of Arab thought that matches that of Hobbes on Western thought.

IBN KHALDUN

Ibn Khaldun lived in Tunis two and a half centuries before Hobbes. His family was of Yemeni descent and had settled in Seville during the Arab expansion. It had later left modern-day Spain after Ferdinand II took Seville in 1248, and had settled in Tunis. Ibn Khaldun was concerned with explaining why a once thriving Muslim Arab empire had declined and collapsed. He was writing after the Reconquista, the Crusades, and the Seljuk, Ottoman, and Mongol invasions had cumulated to bring down the prestigious Arab empire. His concern remained very close to Hobbes's because he sought to understand how civilization (*umran*) could be achieved, where men could live in peace and the arts and sciences flourish (see Ibn Khaldun 1936, 1986*a*, *b*).

Ibn Khaldun's reasoning resembles Hobbes's very closely in its first stages. He considered that men could not live without a society, for an isolated individual could not manage to feed, clothe, or shelter himself unless he cooperated with others (1934: 86) He viewed mutual help among humans as essential for their survival in spite of the fact that their very nature propelled them, as it did animals, to wage war against each other (ibid. 87). This mutual help among humans is what Ibn Khaldun called civilization and what he sought to study.

The gathering of men within a society having been accomplished, as we indicated, and the human species having populated the world, a new need is felt, that of a powerful control that would protect them one from another; for man as an animal, is prone by his nature to hostility and violence. Another means is thus absolutely necessary to prevent these mutual aggressions. No moderator is found among the other animal species, because they have far less perception and inspiration than men; so the moderator needs to belong to the human species and he needs to have a sufficiently strong hand, a sufficiently strong power and authority to prevent men from attacking each other. That is what constitutes sovereignty. We see, from these observations, that sovereignty is an institution that is peculiar to man, that is coherent with his nature and that he cannot do without. (Ibn Khaldun 1934: 89, trans. J. Trottier)

Such thoughts would be very much at home in Hobbes's *Leviathan*. Over two centuries before Hobbes, Ibn Khaldun was convinced of the role played by rationality in building this sovereign power. He disagreed with the philosophers who believed that the authority capable of controlling men could come only from God. He used as a proof the fact that humans had survived as a society long before prophets delivered their message and the fact that pagans were much more numerous than the People of the Book and they also had their dynasties and civilizations (ibid. 90).

Ibn Khaldun disagrees with Hobbes, however, when he considers communitarian forms of social organizations. He perceived such ties as necessary, rather than as merely a hindrance. He valued them as a source of what we would nowadays call 'social capital'. Whereas Hobbes perceived them as a hindrance that must disappear for the state to arise and establish peace, Ibn Khaldun regarded them as essential for the survival of the state. He perceived them as founding the power of the state so long as the latter could manage its relations with them.

Ibn Khaldun devoted much attention to *asabiah*, imperfectly translated as 'tribal spirit', the link bonding an individual to his tribe and family, which enabled survival in the desert. It is what caused people to help each other. For *asabiah* to be strong and efficient, family lineage had to be supported by affection that unites the hearts (ibid. 271). Ibn Khaldun describes in detail the structure of what is nowadays called segmentary nationalism and considers that these community ties (*asabiah*) ensure respect and obedience to a family. He defines *asabiah* as 'the feeling that propels one to resist, to repel the enemy, to protect one's friends, to avenge offences against them' (ibid. 296, trans. Trottier)

Ibn Khaldun had a cyclical vision of history. Writing before the days of political correctness, he thought that half-savage people who live in a harsh desert environment developed strong community ties to survive, which is what allowed them to conquer territories and riches. However this very conquest of riches would lead the following generations to grow up in ease and luxury. *Asabiah* would wither away as it was no longer necessary for survival. (ibid. 275–91). The decline of *asabiah* would slowly lead to the loss of sovereignty.

The forfeiture of social capital via the withering away of communitarian solidarities was regarded as the reason for a state's downfall, whereas

150 years later it would be regarded as the reason for the emergence of the state within Hobbes's work. Interestingly, both Hobbes and Ibn Khaldun considered their theories to have universal value. They both used historical examples to support their ideas. Drawing upon what material was available, Hobbes used European whereas Ibn Khaldun used North African and Middle Eastern examples. Ibn Khaldun illustrates his reasoning on *asabiah* with the case of the Jewish people led into the desert by Moses. They had been brought up in the luxurious environment of Egyptian cities and refused to attack the Syrians. A new generation had to be born and brought up in the desert before they had enough solidarity (*asabiah*) to accomplish God's will, transmitted via Moses, to attack and vanquish (ibid. 296).

In spite of their universal pretensions, both authors seem to have essentially succeeded in explaining culture-specific and region-specific political developments. Ibn Khaldun's ideas are quite illuminating when examining the history of Jordan. Winston Churchill claimed to have created Transjordan 'with the stroke of a pen on a Sunday afternoon'. Yet, the country would not have persisted so long under so few kings had they not respected tribal and community ties and capitalized on them. First, King Abdallah and then his grandson King Hussein had surprisingly long reigns and managed to keep power by maintaining a careful network of ties with all the various communities that make up modern-day Jordan (Gubser 1983). Throughout his life often portrayed as on the verge of losing power, King Hussein died in 1999, the longest-lasting head of state in the region. Western political analysts had viewed the segmentary nationalism in Jordan through Hobbes's eyes and perceived it only as a threat. They had failed to notice how *asabiah* could be harnessed by a state leader and could maintain him in power.

HOBBES, IBN KHALDUN, WATER, AND POWER

A re-examination of the water-war and water-peace literature reveals that we have often used a Hobbesian prism to analyse a Khaldunian world. Researchers have much too often overestimated the state's role in spelling out the rules governing water management. All states put out legislation concerning water that follows similar trends, as illustrated by Stefano Burchi in Ch. 6. Such widely different states as Kyrgyzstan, Uruguay, Morocco, and Indonesia increasingly incorporate within their water legislation principles that are promoted by the World Bank, the FAO, and other such institutions. These are intergovernmental institutions made up of member states. They unavoidably create a state-empowering environment and often neglect to take into account the social capital generated by communitarian institutions involved in the management of water.

A Hobbesian slant may lead to privatization measures, for example, that target the management of water within a public property regime. The companies winning the management contracts may often only discover by experience the working rules in force. Water is often managed in fact according to a communitarian property regime which has remained invisible because a Hobbesian prism did not allow social actors other than the state to be identified and observed.

The social capital that Ibn Khaldun would have called *asabiah* allows social actors to invest network forms of organization quite successfully. Modern networks are not formed on the basis of lineage or family relations, but Ibn Khaldun's definition of *asabiah* does not restrict itself to family ties. It describes the strong feelings, the community ties that will motivate the members to protect and fight for each other. Such feelings can nourish an imagined community spread around the globe and united by a common ideology, whether it be antiglobalization or Islamism. As opposed to a state, this community is not necessarily anchored territorially within state borders. In an era of globalization where network forms of social organization prove quite successful at mobilizing means of violence, a Khaldunian prism seems useful to investigate conflicts over water. It allows us to recognize the variety of social actors who compete for water and the various drives that could allow them to mobilize violence and enter into conflict.

The existing body of literature on water and conflict in the Middle East has focused on states while ignoring the myriad other institutions that participated in controlling water management. Research shows the great importance of these social actors in the Middle East (Oujahou 1985; Trottier 1999, 2000). They generally manage water according to communal property regimes. Many conflicts arise within these institutions and between them and state institutions. This leads to an intricate array of relations of co-operation and competition among previously unresearched social actors.

Re-examining critically the four basic assumptions underlying both the water-war and the water-peace discourse is worthwhile. First, states were portrayed as rational actors that decide whether or not to go to war on the basis of a rational analysis, and, second, sovereignty over war was portrayed as a national interest. Such hypotheses, once suitably amended, are in agreement with most of the theories explaining war that have been generated within the main Western ideologies.

A theory is a set of statements that are internally consistent, a set of causal relations that are assumed as valid, that allow us to explain or predict specific events, and are open to empirical testing. A theory explaining what can lead a state to wage war necessarily derives from its creators' ideology. It determines which questions are deemed relevant and which are not. Historians establish the accuracy of facts and events that lead to war, but the choice of facts they pay attention to is determined by the theory they adhere to. And this theory springs from the ideology they subscribe to.

The hypothesis of states waging wars for water happened to be coherent with the clusters of theories of war generated by the three main Western ideological traditions: conservatism, liberalism, and radicalism (Nelson and Olin 1979). Theories of war generated within the conservative ideology all embrace the pessimistic outlook on human nature that is well expressed by Hobbes. Considering a strong domestic order and a strong leadership is crucial, conservative theories stress the need for a clearly defined international hierarchy or, at least, a well-established balance of international power. The 'realist' school of American foreign policy was a proponent of such of vision (Morgenthau 1973). Henry Kissinger (1994) built on it and maintained that all foreign policy begins with security and the need for an international balance of power.

Conservative theories of war precluded the idea that water be managed by other institutions than the centralized state once it was identified as a national interest. Conservative theories of war also view water-sharing agreements as vital. The issue determining whether war would break out was a correct allocation of water to every riparian that would allow the central government to keep its domestic society in order. Such an allocation did not need to be equal. It only needed to satisfy the needs of every government when facing its potentially troublesome population.

Much literature concerning water wars relies on these hypotheses and causal relations (Naff and Matson 1984; El Musa 1996). These do not survive scrutiny, however, in much of the developing world where the state does not spell out the rules governing use, access to, and allocation of water.

Theories of war generated within the liberal ideology have often provided the implicit blueprint for much of the water-war and water-peace literature. Nelson and Olin (1979) identify three main clusters of theories of war within the liberal ideology. A first cluster, the social-psychological theories, proposes that liberal nation-states behave rationally and cooperatively unless their leaders and citizens fall victim to misperceptions and stereotypes. A second cluster, the structural-functionalist theories, spurred by Talcott Parsons, stresses the existence of mutually beneficial interchanges among interdependent parts. Frustration is generated when needs are not met and this can lead to aggression. Conflict can only occur in a disequilibrated social system that generates such frustration. Such a structural-functionalist theory lies behind the fears generated by Falkenmark's water-stress indicators. It also lies behind Wolf's assertion that water scarcity leads to international cooperation. This type of theory later stimulated the emergence of regime theories. International regimes were defined as sets of principles, norms, rules, and decision-making procedures around which actor expectations converge in a given issue-area (Krasner 1991). Adepts of regime theories thus maintained that states cooperated because they shared norms and values, not because of any power relation among them. Unsurprisingly, such a theory proved most popular in the United States at a time when it was the most powerful state in the world.

The third cluster of theories of war generated by the liberal ideology, group conflict, emphasizes the continuous struggle among groups and among nation-states for limited resources. Such theories acknowledge the existence of coercive relationships and the domination of some groups over others. Although this set of theories showed the greatest potential to recognize the various groups involved in the competition for deciding on how water would be used, allocated and accessed, such was the overwhelming bias in favour of the strong, centralized state that the debate essentially revolved around whether multipolarity or bipolarity generated the greatest risk of war.

Finally, the theories generated within the radical ideology generally rejected both the conservative view that man is innately aggressive and the liberal view that events can be explained in psychological terms or in terms of a breakdown of accommodating structures or balanced systems. Radical theories consider that events are only intelligible in terms of social structures that lie beyond the immediate control of human actors. Thus Lenin built on Hobson's work to argue that capitalism inevitably led to imperialism and to war as capitalist nations were engaged in ruthless competition in their continuous search for cheap raw materials, and markets for commodities, excess capital, and cheap labour. All that was needed to churn out a theory of water war within the radical ideology was to classify water as a precious natural resource. The concern with quantitative analysis of water resources within various arid states certainly fuelled such a classification.

CONCLUSION

Both the water-war and the water-peace discourse emerged within existing theoretical frameworks that were rarely acknowledged yet implicitly structured the researchers' investigations. They severely limited the issue definitions, the choice of factors that were deemed relevant. They led the scientific community to argue for and against water wars for which examples can hardly be found. Significantly enough, Israeli authors, such as Micha Bar, often tended to use regime theories, which downplay power relations, while Palestinian and Arab authors have tended to favour either radical theories, as Ali Hassan Dawod Anbar does, or conservative theories, as El Musa implies in his work (Dawod Anbar 1983; El Musa 1996; Bar 1999). The manner in which they framed the issues and the causal relations they identified thus reflected the ideologies that best suited their identification with either the more powerful or the weaker player.

Water competitions and conflicts are indeed very present and their urgency in much of the developing world is hardly deniable. They pit customary institutions that manage water according to communitarian property regimes against states and private companies as well as against each other. The myriad of interactions such conflicts and competitions give rise to need to be studied in a multiscalar fashion. Existing theories have not equipped us to examine the coming water conflicts because they were often state centric and downplayed the importance of customary and informal institutions, relegating the communal property regimes to the rank of historical vestiges.

Now that network forms of organization have demonstrated their potential for gathering means of violence, a new investigative framework needs to be designed that will allow us to explore conflict dynamics beyond interstate war. We should no longer let a Hobbesian prism obscure a Khaldunian reality at a time when *asabiah* plays a major role both in communitarian and network forms of organization.

REFERENCES

Allan, J. A., 'Substitutes for water are being found in the Middle East and North Africa', *Geojournal*, 28/3 (1992), 375–85.
—— 'Watersheds and problemsheds: explaining the absence of armed conflict over water in the Middle East', *Middle East Review of International Affairs*, 2/1 (1998), 1–4.
Amery, H. A., 'Water, war and peace in the Middle East: comments on Peter Beaumont', *The Arab World Geographer*, 4/1 (2001), 49–52.
Annan, K., Question and answer session after statement (SG/SM/7742) at the Federation of Indian Chambers of Commerce and Industry, New Delhi, 15 March 2001.
Badie, B., *L'État importe. Essai sur l'occidentalisation de l'ordre politique* (Paris: Fayard, 1992).
Barnett, Michael N., *Dialogues in Arab Politics: Negotiations in Regional Order* (New York: Columbia University Press, 1998).
Bar, M., Ph.D. thesis, Hebrew University, Jerusalem, 1999.
Baskin, Gershon, 'The clash over water: an attempt at demystification,' *Palestine-Israel Journal*, 3 (1994), 27–35.
Beaumont, P., 'The myth of water wars and the future of irrigated agriculture in the Middle East', *International Journal of Water Resources Development*, 10 (1994), 9–21.
—— 'Water and armed conflict in the Middle East—fantasy or reality?', in N. P. Gleditsch (ed.), *Conflict and the Environment*, International Peace Research Institute (Dordrecht: Kluwer, 1997).
Benvenisti, Eyal, and Gvirtzman, Haim, 'Harnessing international law to determine Israeli-Palestinian water rights: the Mountain Aquifers', *Natural Resources Journal*, 33/3 (1993), 543–67.
Beshorner, Natasha, 'L'Eau et le processus de paix israélo-arabe', *Politique étrangère*, 4 (1992), 837–55.
Biswas, A. K. (ed.), *International Waters of the Middle East From Euphrates-Tigris to Nile* (Bombay: Oxford University Press, 1994).

Bulloch J., and Darwish, A., *Water Wars: Coming Conflicts in the Middle East* (London: Victor Gollancz, 1993).

Caponera, D. A., *Les Principes du droit et de l'administration des eaux. Droit interne et droit, international* (Paris: Éditions Johanet, 2000).

Dawod Anbar, A. H., 'Socioeconomic aspects of the East Ghor Canal Project, Jordan', Ph.D. thesis, Faculty of Science (Geography), University of Southampton, 1983.

El Musa, S. S., *Negotiating Water: Israel and the Palestinians, A Final Status Issues Paper* (Washington: Institute for Palestine Studies, 1996).

—— 'Hydro-Orientalism: Who is Afraid of Native Experts?', *The Arab World Geographer*, 4/1 (2001), 52–4.

Falkenmark, M., *Hydrological Phenomena in Geosphere–Biosphere Interactions: Outlooks to Past, Present and Future* (Wallingford: International Association of Hydrological Sciences, in co-operation with the International Institute for Hydraulic and Environmental Engineering, 1989).

Finnemore, M., *National Interests in International Society* (Ithaca, NY: Cornell University Press, 1996).

Fukuyama, F., *The End of History?* (Washington: Irving Kristol, 1989).

Giddens, A., *The Consequences of Modernity* (Cambridge: Polity, 1992).

Gubser, P., *Jordan: Crossroads of Middle Eastern Events* (Boulder: Westview, 1983).

Hardin, G., 'The Tragedy of the Commons', *Science*, 162/3859 (1968), 1243–8.

Hobbes, T., *Leviathan* (London: Penguin Classics, 1985).

Homer-Dixon, T., 'On the threshold: environmental changes as causes of acute conflict', *International Security*, 16/2 (1991), 76–116.

Ibn Khaldun, *Les Prolégomènes*, trans. into French and comm. by M. De Slane, pt 1 (Paris: Librairie orientaliste Paul Gauthner, 1934).

—— *Les Prolégomènes*, trans. into French and comm. M. De Slane, p. 2 (Paris: Librairie orientaliste Paul Gauthner, 1936).

—— *Peuples et nations du monde*, trans. from Arabic by Abdesselam Cheddadi, p. 1 (Paris: La Bibliotheque arabe, Sindbad, 1986a).

—— *Peuples et nations du monde*, trans. from Arabic by Abdesselam Cheddadi, p. 2 (Paris: La Bibliotheque arabe, Sindbad, 1986b).

Katzenstein, Peter, *Culture of National Security: Norms and Identity in World Politics* (New York: Columbia University Press, 1996).

Kissinger, H. A., *Diplomacy* (New York: Simon & Schuster, 1994).

Krasner, S. D. (ed.), *International Regimes* (Ithaca, NY: Cornell University Press, 1991).

Lipchin, Clive, 'Public perceptions and attitudes toward water use in Israel: A multilevel analysis', PhD. thesis, Michigan State University, 2003.

Lowi, Miriam, R, *Water and Power. The Politics of a Scarce Resource in the Jordan River Basin*, Cambridge Middle East Library, 31 (Cambridge, Mass.: Cambridge University Press, 1993a).

—— 'Bridging the divide: transboundary resource disputes and the case of West Bank water', *International Security*, 18/1 (1993b), 113–38.

Macpherson, C. B., *The Political Theory of Possessive Individualism, Hobbes to Locke* (Oxford: Clarendon, 1962).

Medzini, A., 'The River Jordan: the struggle for frontiers, water: 1920–1967', Ph.D. thesis, University of London, London, 1997.

Migdal, J. S., *Strong Societies and Weak States, State–Society Relations and State Capabilities in the Third World* (Princeton, NJ: Princeton University Press, 1988).

—— *State in Society* (Cambridge: Cambridge University Press, 2001).

Ministry of Foreign Affairs, Israel, *Joint Declaration for Keeping the Water Infrastructure out of the Cycle of Violence*, 1 February 2001, paragraph (b).

Morgenthau, H., *Politics Among Nations* (New York: Knopf; 1973).

Naff, T., and Matson, R. C., *Water in the Middle East. Conflict or Cooperation?*, Middle East Research Institute, University of Pennsylvania (Boulder: Westview, 1984).

Nelson, K. L., and Olin, S. C., *Why War? Ideology, Theory, and History* (Berkeley: University of California Press, 1979).

Ohlsson, L., 'The Turning of a Screw', *Gestion equitable, efficiente et durable de l'eau pour le développement agricole et rural en Afrique sub-saharienne et dans les Caraibes*, Rapport de synthèse final du séminaire du centre technique agricole, Wageningen, 20–5 Sept. 1999*a*.

——*Environment, Scarcity and Conflict: A Study of Malthusian Concerns*, Department of Peace and Development Research (Göteburg: University of Göteburg, 1999*b*).

Oujahou, L., 'Éspace hydraulique et société. Les systèmes d'irrigation dans la vallée de Dra Moyen (Maroc)', Ph.D. thesis, Université Paul Valery, Montpellier III, 1985.

Postel, Sandra, *Pillar of Sand. Can the Irrigation Miracle Last?* (New York: Norton, 1999).

Ruf, Thierry, 'Sept siècles d'histoire des droits d'eau et des institutions communautaires dans les canaux de Prades (Pyrénées-Orientales)', in O. Aubriot and G. Jolly (eds.), *Histoire et sociéte rurale* (Montpellier: n. pub., 2001).

Soffer, A., *Rivers of Fire: The Conflict over Water in the Middle East* (Oxford: Rowman & Littlefield, 1999).

Starr, J. R., 'Water Wars', *Foreign Policy*, 82 (1991), 17–36.

Starr, J. R., and Stoll, D. C. (eds.), *The Politics of Scarcity, Water in the Middle East*, Center for Strategic and International Studies, Westview Special Studies on the Middle East (Boulder: Westview, 1988).

Trottier, J., *Hydropolitics in the West Bank and Gaza Strip* (Jerusalem: PASSIA, 1999).

—— 'Water and the challenge of Palestinian institution building', *Journal of Palestine Studies*, 19:2/114 (2000), 35–50.

Wendt Alexander, 'Constructing international politics', *International Security*, 20 (1995), 71–81.

Wolf, A., 'Conflict and cooperation along international waterways', *Water Policy*, 1/2 (1998), 251–65.

—— '"Water wars" and water reality: conflict and cooperation along international waterways', in S. Lonergan (ed.), *Environmental Change, Adaptation and Human Security* (Dordrecht: Kluwer Academic, 1999).

8

Water and Development:
A Southern African Perspective

Peter Ashton

INTRODUCTION

Water is acknowledged as the most indispensable of all natural resources, and neither biological diversity nor social and economic development can be sustained in its absence (Hudson 1996; Ashton 2002). Every country faces a similar challenge, namely, providing sufficient water to meet the escalating needs of expanding populations while continuing to ensure that the available resources are used equitably and efficiently (Biswas 1993; Gleick 1998; Ashton and Haasbroek 2002). Increasing rates of industrialization, urbanization, and mechanization aggravate the pressures imposed by population growth, while increasing rates of utilization and pollution place further demands on dwindling resources (Falkenmark 1994, 1999; Rosegrant 1997; Gleick 1998; Ashton 2002). This situation is especially serious in arid regions where water scarcity hinders social and economic development and is linked closely to the prevalence of poverty, hunger, and disease (Falkenmark 1989; Gleick 2000; Ashton 2002).

In southern Africa, water resources are unevenly distributed in both geographical extent and time, and large areas of the region regularly experience prolonged and extreme droughts. Ironically, these droughts are often 'relieved' by equally extreme flood events (Christie and Hanlon 2001). Whilst the availability of water resources is naturally variable and often unpredictable, there is also compelling, though as yet unverified, evidence that projected trends in global climate change could worsen this situation (Ashton 2002). Falkenmark (1989) noted that several African countries had approached or would soon pass the point indicating severe water stress or water deficit, and that this could hinder further development in these countries. Recent estimates suggest that more African countries will exceed the limits of their economically usable, land-based water resources before the year 2025 (Ashton 2002). These disturbing statistics emphasize the urgent need to find sustainable solutions to the problem of ensuring secure and adequate water supplies for all countries in the region.

The consequences of social and political dispensations imposed by previous colonial and apartheid administrations in southern Africa are reflected in the disparate levels of social, economic, and political development attained by these countries. These unequal levels of development have been accompanied by differing levels of need for water, further complicating the search for equitable and sustainable solutions to water supply problems (Ashton 2000, 2002). Whilst equitable access to water to sustain basic human needs is a fundamental and indisputable right of all peoples (Gleick 1999), there is a growing realization that achievement of this ideal requires countries to harmonize their approaches to water management and utilization at national and regional scales (Ashton 2002). Where water allocation and distribution priorities in each country previously supported national development objectives, greater emphasis must now be placed on measures designed to ensure that the region's scarce water resources are used to derive the maximum long-term benefits for all. Clearly, water resource management must be cautious and judicious if this goal is to be achieved and all sectors of society are to have equitable access to the available resources. This is particularly important in shared river basins where a country's water resource management strategies should be aligned with those of its neighbours if peace and prosperity are to be maintained and conflict avoided (Pallett 1997; Ashton 2000, 2002; Ashton and Haasbroek 2002; Heyns 2002).

This brief overview highlights the pressing need to re-examine the availability of water resources in southern Africa and assess the likely trajectories of change in the demand for water that will occur as populations and societies continue to grow and develop in the future. In turn, this will provide a framework to identify those water resource management strategies that can contribute to sustainable social and economic development across the region. The likely success or failure of different strategies will determine whether or not regional development goals can be achieved and the looming potential for conflict over water resources can be avoided (Ashton 2002).

AVAILABILITY OF WATER IN SOUTHERN AFRICA

Geographical considerations and climatic controls

Geographically, the African continent occupies approximately equal latitudes either side of the Equator, with some 57 per cent of the continent's land mass located in the northern hemisphere (SARDC 1996). During a normal annual cycle, the equatorial regions of Africa receive considerably more rainfall than the southern and northern portions of the continent. The drier areas also experience greater variability in year-to-year rainfall, more extreme air temperatures, and higher rates of evaporation. In combination, these features reduce surface water flows and provide little recharge to groundwater aquifers. In addition, recurring cycles of the El Niño southern oscillation (ENSO) phenomena have been linked to the periodic occurrence of severe droughts interspersed with exceptional flooding events across large areas of southern Africa (SARDC 1996; Christie and Hanlon 2001). The devastating recent floods of 1996, 1998, 2000, and 2001 in Mozambique revealed several shortcomings in existing water resource management and disaster mitigation strategies throughout southern Africa (Christie and Hanlon 2001). This has raised concerns that new and improved management approaches are needed to deal with flood events and their social and economic repercussions in the region.

Typically, the average quantity of surface run-off in southern African countries ranges from 20 per cent of mean annual rainfall in wetter areas to zero in the deserts (Ashton 2002). This is in marked contrast to many equatorial African countries where surface run-off regularly exceeds 30 per cent of mean annual rainfall. The exceptionally high river flows that caused severe flooding in Mozambique and parts of Botswana, South Africa, and Zimbabwe during 2001 were triggered by prolonged periods of above-average rainfall across the subcontinent, following an earlier year of higher than normal rainfalls (Christie and Hanlon 2001).

Across southern Africa, the long-term geometric (spatial) average annual rainfall amounts to 948mm, 45 per cent higher than Africa's average of 652mm (Table 8.1). While high average annual rainfalls in Angola, DRC, Malawi, and Zambia combine to raise the SADC average rainfall above the world average of 860mm, large areas of Botswana, Namibia, and South Africa are extremely dry and receive substantially less than 400mm of rainfall per year (SARDC 1996; Ashton 2002; Heyns 2002). Southern Africa has few primary aquifers or high-yielding geological formations and most groundwater occurs in small, scattered volumes in secondary, fractured rock aquifers (Basson *et al.* 1997).

The general pattern of climatic characteristics across southern Africa has resulted in a striking absence of perennial rivers and lakes in the south-western and north-eastern portions of this region (Ashton 2000). Rainfalls in the driest areas are both erratic and unpredictable; consequently, river flows are usually ephemeral or episodic, interspersed by dry periods that may last for several years. These features are also reflected in the average quantities of annually renewable surface-and groundwater resources that are available in southern African countries (Table 8.1). Importantly, the estimates shown in Table 8.1 do not include volumes of water that are received from, or that flow to, countries that share river systems (Ashton 2002).

TABLE 8.1. *Comparison of the area, mean annual rainfall, and volumes of annually renewable surface- and groundwater available in the twelve SADC mainland states, with the totals for the SADC states and Africa as a whole*

Country	Area (km²)	Mean annual rainfall (mm)	Available surface water (km³)	Available groundwater (km³)	Total water available (km³)
Angola	1,246,700	1,050	184.0	21.0	205.0
Botswana	581,730	400	0.4	1.2	1.6
DRC	2,344,885	1,534	900.0	119.0	1,019.0
Lesotho	30,355	760	4.7	0.5	5.2
Malawi	118,484	1,014	16.1	1.4	17.5
Mozambique	801,590	969	100.0	17.0	117.0
Namibia	824,290	254	1.0	1.7	2.7
South Africa	1,221,000	497	48.0	5.0	53.0
Swaziland	17,363	788	2.6	0.2	2.8
Tanzania	945,087	937	50.0	30.0	80.0
Zambia	752,614	1,011	80.0	47.0	127.0
Zimbabwe	390,760	652	14.0	1.5	15.5
SADC TOTAL	9,274,858	948	1,400.8	245.5	1,646.3
AFRICA TOTAL	29,491,327	652	3,716.3	599.7	4,316.0
SADC as % of Africa	31.5	145.4	37.7	40.9	38.1

Sources: CIA (2000); FAO (2000*b*); Ashton and Ramasar (2002).

The uneven and seasonally variable distribution of water across southern Africa forced water resource managers to develop large storage reservoirs to provide adequate supplies of water during dry periods. Over time, the growing demands for water from the domestic, industrial, and agricultural sectors have had to be met by the construction of new reservoirs. Where the demands for water were located far from the available supplies, these had to be met by means of water transfer schemes (Basson *et al.* 1997). The need to construct increasing numbers of storage reservoirs to meet society's demands for water has attracted considerable attention in recent years and alternative approaches may need to be considered in future (WCD 2000).

Many households and small communities in the drier regions of Botswana, Mozambique, Namibia, South Africa, and Zimbabwe have to rely on groundwater supplies obtained from boreholes and hand-dug wells (Heyns *et al.* 1998). These communities are particularly vulnerable to water shortages caused by droughts or over-exploitation, as well as contamination that may arise from inappropriate land-use practices (Basson *et al.* 1997).

Despite the limited data available on river flows in many southern African countries, the long-term average quantity of water available within a country appears to remain more or less constant, though several cycles of variability are clearly evident (Conley 1995). Years of above-average rainfall provide short-term relief in the form of additional water, whilst lower rainfalls cause drought conditions that accentuate water shortages. Numerous authors (e.g. Falkenmark 1989; Conley 1995; Ohlsson 1995; Gleick 1998; Ashton 2002) have repeatedly stressed the finite and vulnerable nature of Africa's freshwater resources and this view is slowly gaining widespread acceptance.

In addition, there is now far greater awareness amongst southern African water resource managers that it is essential for countries that share river basins to collaborate more closely and co-ordinate their activities. It is to be hoped that this will also improve the approaches and technologies used for disaster management when dealing with floods and droughts (Christie and Hanlon 2001).

The level of conventional water resource utilization in many southern African countries is very high and new approaches will be needed to stretch the limited water supplies available to meet projected demands for water (Basson *et al.* 1997; Ashton and Haasbroek 2002; Heyns 2002). Significant research into new technologies and sources of supply has resulted in the development and evaluation of a number of innovative concepts and methodologies, as well as novel adaptations to existing approaches. These concepts and methodologies include: integration of surface water transfers into a national water grid, transfers of untapped surface water resources from one country to another, exploitation of deep groundwater and the use of aquifers for storage of surplus water, atmospheric water (fog and cloud) harvesting, iceberg water utilization, desalination, and direct use of seawater. Whilst some of these options are still theoretical and unproven, others have reached different stages of practical testing and implementation (Smakhtin *et al.* 2001). In Namibia, for example, growing water shortages in coastal communities located in the Namib Desert will in future have to be met by routine use of desalinated seawater (Heyns 2002).

Water requirements in southern Africa

There is a growing awareness that increased population numbers and improved quality of life (i.e. 'development' in its widest sense) contribute to a continual increase in the quantity of water needed to meet society's needs. Here, both the consumptive and non-consumptive uses of water are considered together when estimating the water needs of society. The consequences

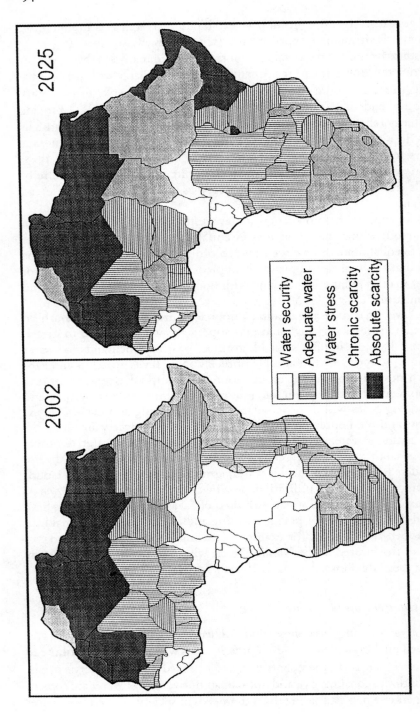

FIG. 8.1 Comparison of the quantity of water available per person for each African country in 2000 and 2025, coded according to the 'water crowding' limits proposed by Falkenmark (1986, 1989). *Sources*: CIA (2000), FAO 2000b, and redrawn from Ashton (2002).

of this process translate into an inevitable reduction in the quantity or fraction of a country's water resources that remain available for use per person
(Falkenmark 1986; Gleick 1998; Ashton 2000). The implications of this
trend for every African country have been coded according to the 'water
crowding limits' proposed by Falkenmark (1986, 1989) and are shown in
Fig. 8.1. The projected reduction in per capita availability of water is accompanied by an escalating trend in the degradation of its quality, and represents a serious threat to the flows of various goods and services required by
developing societies across Africa (FAO 2000*a*; Ashton 2002). Fortunately,
the impending water crisis in southern Africa has helped to enhance public
recognition of the values generated by or linked to water, and has also
improved our understanding of the significant social and economic values
embodied in the many ecosystem services that depend on water (Falkenmark 1999; FAO 2000*a*; Ashton 2002).

The available information on current and projected quantities of water
available in African countries clearly shows that important changes will
occur during the foreseeable future (Fig. 8.1), though the anticipated patterns of change are not uniform across the continent. Several fortunate
countries will remain in a state of 'water abundance' for many years to come,
whilst others will experience increasing levels of 'water scarcity' and 'water
deficit' (Falkenmark 1986). The implications of the anticipated changes for
water resources in southern Africa are shown in Table 8.2, where com-
parisons are made for the years 2000 and 2025, taking into account projected
rates of population growth.

A brief explanation of the data in Table 8.2 helps to highlight the scale of
the problem. In 2000, three SADC counties (Angola, DRC, and Zambia)
had 'water abundance' whilst a further five countries (Lesotho, Malawi,
Mozambique, Swaziland, and Tanzania) had 'adequate' water supplies.

TABLE 8.2. *Comparison of the proportional changes in area, total
population and proportion of water available within each of three classes of
water availability (surface-plus groundwater), for 2000 and 2025 (%)*

Countries with water:	Year 2000 (population 201m.)			Year 2025 (population 331m.)		
	Area	Population	Water	Area	Population	Water
Abundance	67.4	70.3	95.6	55.5	54.2	89.2
Stress	17.4	28.0	4.1	16.2	29.6	7.3
Scarcity	15.2	1.7	0.3	28.3	16.2	3.5

Note: Data refer only to the twelve SADC countries on the African continent.

Sources: CIA (2000); FAO (2000*b*).

These eight countries comprise some 67 per cent of the area of the SADC region, are home to 70 per cent of the region's population of 201m. people, and contain some 95 per cent of the total renewable surface- and ground-water resources. In contrast, 30 per cent of the SADC population occupy the remaining 32 per cent of the region and have access to a mere 4 per cent of the region's renewable water resources, experiencing either 'water stress' or 'water scarcity' (Ashton 2000).

By the year 2025, it is anticipated that no SADC country will have 'water abundance', though four countries (Angola, DRC, Mozambique, and Zambia) will now have 'adequate' water supplies. The total area of these four SADC countries comprises 55 per cent of the region, housing 54 per cent of the (expanded) population, and containing 89 per cent of the region's renewable water resources (Table 8.2). In contrast, the combined area of the remaining eight SADC countries will be classed as having either 'water stress' or 'water scarcity', and will comprise 44 per cent of the region. Whilst these countries will house some 45 per cent of the region's population, they will contain only 11 per cent of the region's renewable water resources in 2025 (Ashton 2002).

Importantly, these numerical forecasts assume no change in the total quantity of water available within each country, and the SADC population is projected to increase by some 65 per cent to 331m. in 2025 (Table 8.2). However, such population growth estimates must be treated with caution. Given the widespread incidence of waterborne diseases such as malaria and cholera (SARDC 1996; Pallett 1997), together with the enormous implications of the African HIV/Aids pandemic (UNAIDS 2000; Whiteside and Sunter 2000; Ashton and Ramasar 2002), and declining levels of social and economic stability (Biswas 1993; FAO 2000*a*), these forecasts may over-estimate the southern African population in 2025. Conversely, if a cure for HIV/Aids is found, population growth rates may increase in future (Ashton and Ramasar 2002).

Regional patterns of water use

Broad patterns of (predominantly consumptive) water use by different sectors in the SADC region are shown in Table 8.3. Although comparative data on the total volumes of water used consumptively and non-consump-tively in each country are absent, agricultural water use clearly dominates when compared to proportions used by the domestic and industrial sectors (SADC 2000; Ashton and Ramasar 2002). The high proportion of water used for agriculture suggests that each SADC country relies heavily on food grown within its borders to meet national goals of food security (Pallett 1997). A proportion of the water used for irrigated agriculture is considered to be non-consumptive use because it returns to nearby surface- and ground

TABLE 8.3. *Water use by different sectors in twelve mainland Southern African (SADC) countries in 1998*

Country	Proportion of water used by different sectors (%)		
	Agriculture	Industry	Domestic
Angola	76	10	14
Botswana	48	20	32
DRC	23	16	61
Lesotho	56	22	22
Malawi	86	3	10
Mozambique	89	2	9
Namibia	68	3	29
South Africa	62	21	17
Swaziland	71	8	21
Tanzania	89	2	9
Zambia	77	7	16
Zimbabwe	79	7	14

Note: Industry sector includes mining activities.

Sources: Gleick (1999); WRI (2000); Ashton and Ramasar (2002).

water systems through sub-surface return flows. However, the quality of this water is usually poor due to its increased content of dissolved salts (Basson *et al.* 1997). Additional, usually non-consumptive, water uses include the provision of sufficient water to sustain the structure and functioning of aquatic ecosystems so that they are able to continue providing the variety of ecological goods and services (e.g. food, flood attenuation, nutrient purification, medicinal plants, aesthetic values, etc.) that are demanded by society (Ashton and Haasbroek 2002).

Another extremely important consideration, and an indirect indicator of the level of development attained by society, is the degree of access to sanitation services and safe, wholesome supplies of water (Table 8.4). As shown by the data presented in the table, southern African countries show wide disparities in their degree of urbanization, ranging from Malawi at 14 per cent to Botswana at 64 per cent. Overall, some 69.1 m. people (34.4 per cent), out of a total SADC population of 201.2m. live in formal urban areas while 132.1m. people (65.6per cent) live in rural areas. This proportion is similar to the average for Africa (Table 8.4).

An examination of the urban populations of southern African countries reveals wide differences in the provision of water supply services, ranging from 17 per cent in Mozambique to 100 per cent in Botswana (Table 8.4).

TABLE 8.4. *Degree of access to safe water supplies and appropriate sanitation systems in the urban and rural populations of twelve SADC countries in southern Africa, compared with the total for the mainland SADC countries and for Africa as a whole*

Country	Population (m.)	Proportion urbanized (%)	Safe water (%)		Sanitation (%)	
			Urban	Rural	Urban	Rural
Angola	12.9	31	69	15	34	8
Botswana	1.6	64	100	91	91	41
DRC	52.0	29	37	23	23	4
Lesotho	2.1	25	65	54	53	36
Malawi	10.8	14	80	32	52	24
Mozambique	19.9	35	47	40	53	15
Namibia	1.7	37	87	42	77	32
South Africa	43.3	49	80	49	79	50
Swaziland	0.9	32	61	44	66	37
Tanzania	33.7	25	67	45	74	42
Zambia	9.2	43	64	27	75	32
Zimbabwe	13.1	43	90	69	90	42
SADC TOTAL	201.2	34	69	39	61	30
AFRICA TOTAL	789.9	35	65	38	58	34

Sources: CIA (2000); FAO (2000*b*); Ashton and Ramasar (2002).

Overall, some 43.1m. people (62 per cent of all urban residents) can access safe water supplies, while the remaining 26.2m. urban residents have no such access. A slightly smaller number (42.3m., 61 per cent) of all urban residents also receive some form of formal sanitation service. Once again, the proportions recorded in southern Africa are similar to those for the African continent as a whole (Table 8.4).

In comparison to their urban counterparts, the much larger rural populations of southern African countries have far lower levels of access to appropriate sanitation services and safe water supplies (Table 8.4). Out of a total rural population of 132.1m., only 46.9m. (36 per cent) can access safe water supplies while 39.3m. people (30 per cent) have appropriate sanitation facilities. Again, wide disparities are noticeable between the levels of services available within the different southern African countries. For example, access to safe rural water supplies ranges from 15 per cent in Angola to 91 per cent in Botswana, while access to appropriate sanitation services ranges from

4 per cent in the Democratic Republic of Congo to 62 per cent in Tanzania (Ashton and Ramasar 2002).

In summary, 90m. (44.7 per cent) of the total southern African population have access to safe water supplies while 81.6m. people (40.5 per cent) have access to appropriate sanitation facilities. Importantly, these figures also show that 111.5m. people (55.3 per cent) have inadequate access to safe water supplies while 120m. people (59.6 per cent) have inadequate access to appropriate sanitation facilities. While these figures reflect the relatively large southern African population that is in need of service provision, the rate at which these services are being provided appears to lag behind the growth in demand (Ashton and Haasbroek 2002).

Existing infrastructure and water resource management systems

Broadly speaking, water resource management requires that a delicate balance be achieved between protecting water resources while simultaneously ensuring that the reasonable demands for water by each water use sector are met in a timely manner (Ashton and Ramasar 2002). The accomplishment of this equilibrium requires the inputs of large numbers of skilled and semi-skilled individuals from a wide variety of economic, social, and technical disciplines. The success of these activities relies heavily on the collection and interpretation of information relating to the availability of water of suitable quality, the geographical and temporal distribution of demands for water, as well as the design, construction, and operation of appropriate water supply and treatment works to meet these demands (Basson *et al.* 1997).

Depending on their size and complexity, water supply schemes and their associated infrastructure can take 3–15 years to commission from the time they were first conceptualized. It is therefore imperative that water resource managers have rapid access to accurate and current information on the demographic distribution of populations and their likely future water demands (SADC 2000; Ashton and Haasbroek 2002). Insufficient supplies of water cause unnecessary hardship and stress, coupled to retarded social and economic growth, whilst over-provision of water leads to wastage, environmental damage, and economic loss (Ashton and Haasbroek 2002).

Modern approaches to water resource management recognize that water resources can be managed effectively and efficiently only when the entire river basin or catchment forms the basic management unit. Furthermore, because surface water and groundwater are inextricably interlinked, they must be considered and managed together as a single resource. These principles form the foundation for integrated water resource management (IWRM), and are rapidly gaining wider acceptance throughout the world (Newson 1992; Shela 1996; Ashton and Haasbroek 2002). Most southern African countries have recognized the need for IWRM approaches and have already drawn up policies, implemented the required legislation, and initiated actions designed to achieve these objectives (Asmal 1998). This is a very promising start though it will still take some time for the full range of benefits of these activities to be realized.

In their ideal form, IWRM approaches to catchment management provide both the guiding philosophy and a practical framework for actions that promote cooperative decision-making and responsible management of water resources. A basic tenet of catchment management is the principle that all water users within a catchment must share responsibility for determining the short-, medium-, and long-term objectives of water resource management, while ensuring that water allocation is both equitable and fair (ibid.).

Many southern African water laws have been modified from their original (colonial) form and now share several common features (SARDC 1996; SADC 2000). Particularly important are those aspects that recognize water as a common good, denote each state as having a custodial responsibility for water, and replace previous situations of 'ownership' of water by individuals with a common 'right to the fair and equitable use of water' (Asmal 1998; Ashton and Haasbroek 2002). Whilst some of the principles contained in these legal systems represent a dramatic departure from previous water law, they now provide a far more equitable basis for water allocation and management (Asmal 1998).

Unfortunately (as shown in Table 8.3), many southern African governments have not been able to provide appropriate levels of water supply and sanitation services to all communities (SARDC 1996). In part, the problems of slow service delivery have been caused by the enormous backlogs in service provision that were inherited from earlier colonial or apartheid governments. These difficulties have been compounded by rapidly ageing water supply infrastructure, shortages of economic resources, and insufficient technical capacity. In an effort to address the critical need to accelerate service delivery, many governments are contemplating, or have already entered into, partnerships with the private sector to achieve the required levels of service provision. Clearly, several different types of partnership option are available, ranging from complete transfer of the relevant assets, infrastructure, and functions to the private sector (i.e. full privatization in its strictest sense), to various forms of agency-type agreements. In the latter, the relevant national government authority retains overall responsibility and accountability and the agency provides the necessary skilled manpower and technical resources to manage the distribution of services. Regrettably, public misunderstandings about the need to recover the costs of service provision have occasionally led to rejection of these public-private partnership arrangements (Basson *et al.* 1997; Heyns *et al.* 1998; Eberhard 1999).

DEVELOPMENT IMPERATIVES AND REGIONAL INITIATIVES

In its most basic sense, the term 'development' implies a systematic change of state from an 'older' or 'lower' state, to a 'newer' or 'improved' state, that is demonstrably different from the original. When reference is made to the level of development of a community or society, it is important to remember that development includes the process of change as well as the changed state that is achieved over time. Clearly, development should be seen as part of a continuum, often consisting of discrete stages, where each stage of development may occur at irregular intervals dictated by the availability of resources (Newson 1992). Unfortunately, the term 'development' is often used inappropriately to describe the growth, expansion or enlargement of an existing situation or system, with little evidence of systematic change or improvement having taken place.

The concept of Sustainable Development first entered the global environmental debate in the 1980s to express the fundamental interdependence between economic development, the natural environment, and people. The most widely accepted definition of sustainable development describes it as 'development that meets the needs of the present without compromising the ability of future generations to meet their needs and aspirations' (WCED 1987). This expression of sustainable development seeks to establish a development path that enhances the quality of life of humans whilst ensuring the viability of the natural systems on which that development depends.

In southern Africa, sustainable development initiatives focus on improving the social and economic equity of the poor majority, the rational management and use of natural resources, the eradication of pervasive poverty, the development of the human population, and the improvement of economic growth (Asmal 1998; SADC 2000). The critical importance of sustainable development in Africa has been emphasized in the context of the New Partnership for Africa's Development (NEPAD) and in Africa's preparations to host the 2002 World Summit on Social Development (SADC 2000).

Among the key questions being asked of African countries is the extent to which Agenda 21 and the various environmental conventions have advanced Africa towards the goal of sustainable development (Shela 1996). This focus of attention is pertinent because, despite its enormous natural resource potential, Africa has the highest proportion of arid and semi-arid areas of any continent, with a substantial number of countries and populations experiencing frequent droughts and being adversely affected by desertification (SARDC 1996). In addition, the continent is plagued by widespread poverty, inequity, and endemic diseases, and supports an array of difficult socio-economic conditions brought about by fluctuating terms of trade and

Peter Ashton

TABLE 8.5. *Selected demographic and economic characteristics of the twelve mainland SADC states*

Country	Population (m.) year 2000	Adult HIV/Aids prevalence (%)	Life expectancy at birth (yrs.)	GDP (US$m.) year 2000	GDP-PPP per person per year (US$/P/Yr)	GDP proportion by sector (%)		
						Agric.	Indus.	Serv.
Angola	12.9	2.78	47	11,600	899	8	71	21
Botswana	1.6	35.80	47	5,700	3,478	5	40	55
DRC	52.0	5.07	51	35,700	686	58	17	25
Lesotho	2.1	23.57	46	4,700	2,180	14	36	50
Malawi	10.8	15.96	40	9,400	872	35	20	45
Mozambique	19.9	13.22	46	18,700	936	33	12	55
Namibia	1.7	20.00	53	7,100	4,083	12	26	62
South Africa	43.3	22.60	55	296,100	6,844	5	39	56
Swaziland	0.9	25.30	40	4,200	4,526	19	43	38
Tanzania	33.7	8.09	48	23,300	690	37	21	42
Zambia	9.2	19.95	41	8,500	925	28	37	35
Zimbabwe	13.1	25.06	45	26,500	2,022	11	14	75
SADC TOTAL	201.2		49	451,500	2,241	14	34	52

Note: On average, adults comprise 50% of country populations in SADC.

Source: FAO 2000*a*; UNAIDS 2000; Ashton and Ramasar 2002.

external indebtedness (Table 8.5). Furthermore, Africa as a whole can be characterized by the heavy reliance of its populations on natural resources for their subsistence, insufficient institutional and legal frameworks, poor governance, a weak infrastructure base, and insufficient scientific, technical, and educational capacity. In combination, these characteristics pose enormous challenges to meeting the objectives of sustainable development (Shela 1996).

A brief review of the demographic and economic data presented in Table 8.5 highlights some of the key concerns in southern African countries. First, the alarmingly high adult prevalence of HIV/Aids has caused a dramatic drop in life expectancies at birth. Adults in the age-group 20–40 years of age are most vulnerable to HIV/Aids (UNAIDS 2000) and, since these individuals are usually the most economically active members of society (Whiteside and Sunter 2000; Williams *et al.* 2000), values for per capita GDP can be expected to decline as infected individuals die (Ashton and Ramasar 2002).

Secondly, the relative strength of each country's economy (as represented by GDP) appears to be linked to an increased contribution of the industrial and service sectors of the economy (Table 8.5). This feature is clearly dependent on the availability of trained and capable staff who, in southern Africa, are those at greatest risk of HIV/Aids infection (Whiteside and Sunter 2000).

HIV/Aids has been recognized as a development crisis that threatens the social, economic, and political fabric of southern African societies and the full consequences of the pandemic have yet to be felt (Ashton and Ramasar 2002). However, there is increasing evidence that concerted actions to address the social and economic roots of people's vulnerability to HIV/Aids are starting to bear fruit and there are signs that the pandemic is beginning to be controlled in some areas (UNAIDS 2000).

The relatively low levels of human development in southern Africa have prompted the heads of the SADC countries to focus their goals for sustainable development on equity issues (Hounsome and Ashton 2001). In terms of the SADC vision for sustainable development (SADC 2000), the region must:

1. Accelerate economic growth with greater equity and self reliance,
2. improve the health, income, and living conditions of the poor majority, and
3. ensure equitable and sustainable use of the environment and natural resources for the benefit of present and future generations.

Importantly, a fourth dimension, namely governance, needs to be added to the three goals identified by southern African countries to ensure sustainable development (Hounsome and Ashton 2001; Fig. 8.2). The global frustration about lack of progress in the implementation of various global treaties and protocols is also experienced at company and local government level (MMSD 2001). A future focus on implementation will depend on the creation of institutions of governance that can meet stated objectives. A key

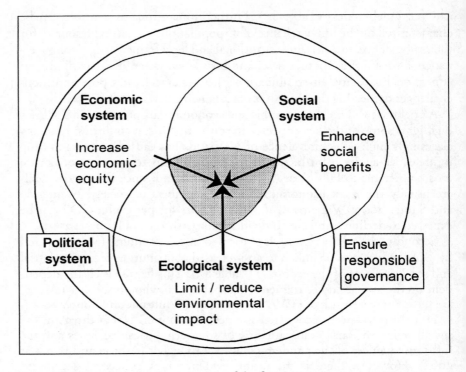

FIG. 8.2 Diagrammatic representation of the four interacting components or systems that comprise sustainable development (shaded). *Source*: Redrawn from Hounsome and Ashton (2001).

part of this process will require stakeholders to engage transparently to promote participation and form partnerships between governments and private individuals or organizations where they are also accountable for their actions and decisions.

In southern Africa, the mining industry has played a pivotal role in the development of infrastructure and in the establishment of manufacturing and other industries (Hounsome and Ashton 2001; MMSD 2001). Although the direct contribution of mining to the economies of SADC countries has declined during the past thirty years, the importance of manufacturing industries based on a wide variety of minerals as raw materials has grown substantially. A brief summary of some of the social and economic contributions currently made by mining to the economies of mainland SADC countries is given in Table 8.6. Importantly, these data exclude associated industries and beneficiation programmes, and represent a conservative estimate of the recent (1999–2000) direct contributions made by the formal mining sector (MMSD 2001).

The data presented in Table 8.6 show that, for southern Africa, mining contributed an average of 10 per cent of the regional total to the per capita GDP for 1999–2000. Whilst the actual contribution for each country varied between 0.5 per cent (for Lesotho) and 52.3 per cent (for Angola), countries with well-developed mining sectors have derived enormous economic benefits (CIA 2000). These have been reflected in additional industrial development as well as improved social facilities and infrastructure (World Bank 1998; SADC 2000; WRI 2001).

Despite prevailing economic pressures on the mining industry, the mining sector within each of the twelve mainland SADC countries employs a highly significant proportion (5.3 per cent) of southern Africa's total available workforce of some 68m. people (Table 8.6). This contrasts sharply with

TABLE 8.6. *Summary data to show the key role of mining in the economies of southern African countries*

Country	GDP in year 2000 (US$M)	Mining contribution to per capita GDP[1] (%)	Mining employment[2] (% of total labour force)	Mining share of total foreign earnings in 1999 [3] (%)
Angola	11,600	52.3	9.0	90.0
Botswana	5,700	38.0	5.5	70.0
DRC	35,700	28.0	4.0	70.0
Lesotho	4,700	0.5	1.0	0.3
Malawi	9,400	0.9	0.4	0.5
Mozambique	18,700	2.0	1.3	2.0
Namibia	7,100	20.0	4.5	48.0
South Africa	296,100	8.0	9.0	28.6
Swaziland	4,200	2.0	3.5	2.0
Tanzania	23,300	2.8	2.7	22.0
Zambia	8,500	12.1	9.0	80.0
Zimbabwe	26,500	8.0	7.0	40.0
SADC TOTAL	451,500	10.0	5.3	30.0

[1] Includes offshore and onshore oil and gas plus diamonds, as well as all other formal land-based mining activities, but excludes all mineral beneficiation contributions. Converted from national GDP data, divided by national population numbers (Table 8.5).
[2] Expressed as a percentage of the total country workforce; does not include artisan miners and linked industries.
[3] Share of total foreign exchange earnings in 1999 that are directly attributable to the sale of minerals and mineral products (excludes allied industries that may beneficiate mineral products).

Sources: CIA (2000); FAO (2000*b*); Hounsome and Ashton (2001).

the estimates for levels of unemployment in each country (CIA 2000) and emphasizes the mining sector's importance as a source of employment. However, these employment figures do not take account of the large numbers of migrant mine workers, nor do they include 'artisan' and 'informal' miners, whose numbers appear to vary between 200,000 and 850,000 (MMSD 2001).

The economic benefits of mining are also reflected in the contribution to direct foreign exchange earnings in each country (Table 8.6). In particular, Angola, Botswana, DRC, Namibia, South Africa, Tanzania, Zambia and Zimbabwe obtain between 22 per cent (Tanzania) and 90 per cent (Angola) of their foreign exchange directly from mining and mineral exploitation activities. Overall, mining is estimated to have contributed approximately 30 per cent to the direct foreign exchange earnings of SADC countries in 2000 (CIA 2000; Ashton and Hounsome 2001; MMSD 2001).

Importantly, the estimates of foreign earnings derived from mining (Table 8.6) exclude economic contributions made by 'artisan' miners and the 'indirect' earnings of migrant miners who, for example, work in South Africa and remit portions of their salaries to their home country. In Lesotho, for example, the wages of migrant miners are estimated to contribute some 20 per cent to Lesotho's foreign earnings and this source of income is also important for Malawi, Mozambique, Namibia, and Zimbabwe (ibid. 2001). Overall, despite fluctuating market forces, mining and mineral processing activities form the economic cornerstone of most southern African countries and provide the foundations upon which their social and economic development programmes are based (MMSD 2001).

The pressing need of southern African countries to achieve a large measure of 'food security' and to transform many of the adverse implications of landownership practices (the so-called 'land reform process') has driven several changes in the region's agricultural sector. Unfortunately this sector is widely seen to be using the greatest quantities of water and to be the most 'wasteful' user of this precious commodity (SARDC 1996; Basson *et al.* 1997; Falkenmark 1999). In recent years there has been a dramatic increase in the use of new cost-effective irrigation techniques that help to use water more efficiently and effectively. However, this trend is not uniform across the region. Several countries lag behind in the implementation of new technologies and approaches and will require skilled technical assistance if they are to achieve meaningful improvements in water use (SARDC 1996).

WATER RESOURCE MANAGEMENT CHALLENGES

The rapidly growing public recognition of water interdependence in many African countries supports an increasing drive towards cooperative development of water resources in certain areas (Falkenmark 1989; Biswas 1993; Gleick 2000). In southern Africa, recent political developments have been accompanied by a wider acceptance of the need for countries to work together to design and implement joint strategies to protect and manage regional water resources (SADC 1995; SARDC 1996; Pallett 1997; Ashton 2002). Large areas of the sub-continent are arid to semi-arid and the basins of most of the larger perennial rivers are shared between three and eight countries (SARDC 1996; Pallett 1997). The growing demands for water in some parts of southern Africa are fast approaching the limits of exploitation that conventional technologies can provide (Conley 1995; Heyns 1995). In many cases, demands for additional supplies of fresh water will have to be met through the use of unconventional technologies, the exploitation of new or novel sources of fresh water, or through the long-distance transfer of ever-larger quantities of water from regions that have ample supplies (Conley 1995; Heyns 1995, 2002; SARDC 1996; Smakhtin *et al.* 2001; Ashton 2002). One of the most pressing challenges faced by water resource managers is the need to manage water demand.

In normal everyday practice, water resource managers can choose from a wide range of technical, economic, and social interventions to conserve water and manage water demand within a country. Individual options or combinations of options can be selected according to specific objectives, or the perceived urgency or need of the situation. Each approach (technical, social, and economic) can then be applied within appropriate timeframes for each water use sector (Ashton and Haasbroek 2002). The success or failure of each water demand management strategy depends on the commitment of water resource managers to implement the chosen strategies, and the willingness of individual water users to abide by the conditions of each strategy. These measures take time to implement and their consequences also take time to impact on patterns of water use. Ideally, each country needs to adopt a fully integrated approach that combines short- medium-, and long-term considerations (ibid.).

The current reality of southern Africa is one of expanding populations accompanied by escalating urbanization and industrialization, as well as demands for water to redress past social, economic, and political iniquities. National water resource management strategies in southern Africa now recognize water as a 'common good' and not as 'private property' (Asmal 1998). The principles of sustainable resource utilization underpin national water resource management policies and ensure that all aspects of the water cycle are considered within the geographical bounds of a river basin or catchment area (Heyns 1995; SARDC 1996; Basson *et al.* 1997).

The effective, efficient, and integrated management of water resources shared by several countries requires a high degree of trust between the countries, as well as a firm commitment to interstate collaboration and cooperation (Lundqvist 2000; Ashton 2002). These responsibilities are seldom easy to incorporate into the existing institutional structures within each country and many of the policies, priorities, and strategies that are needed extend beyond the boundaries of conventional government departments (Wolf 1999). Experience elsewhere in the world has shown that the establishment of a river basin organization (RBO) that represents the interests of all countries sharing a river basin has the greatest likelihood of success (Lundqvist 2000; van der Zaag, Seyam, and Savenije 2000).

The creation of such an RBO requires each state within the river basin to acknowledge and accept the roles and responsibilities of its partners, whilst committing itself to the maintenance of a spirit of harmony and goodwill amongst its partners (OKACOM 1994; Pallett 1997; Lundqvist 2000; van der Zaag, Seyam, and Savenije 2000). An important part of any such international partnership is the realization that the rights and obligations of each party are mutual and reciprocal, rather than unilateral (Wolf 1999; van der Zaag and Savenije 2000). Inevitably, the basis for any agreement on the quantities of water required by a country will depend on the ability of each country to demonstrate its capability to manage the water resources available in a fair and equitable manner (Ashton 2002).

COPING STRATEGIES AND MANAGEMENT OPTIONS

Most statistics on water exploitation and use refer to the more easily accessible and manageable surface- and groundwater components. An important consideration is the availability of the necessary social, economic, and technical resources needed to exploit the available water to meet society's needs. Collectively, these features have been referred to as 'the coping capability' or 'social adaptive capacity' of a society that enable it to take advantage of the available natural resources (Ohlsson 1995; Turton 1999; Turton and Ohlsson 1999; Ashton 2002; Ashton and Haasbroek 2002). In particular, this capability requires a high degree of human ingenuity, the ability to mobilize economic and technological resources and to adapt and adopt plans, strategies, and tactics that help to promote more effective and efficient use of water (Turton 1999; Ashton 2002). Indeed, there is convincing evidence that those countries (such as Israel) that display a highly developed social adaptive capacity have overcome severe water shortages, whilst other countries (such as Burundi) with less evidence of social adaptive capacity, have been unable to do so (Allen and Karshenas 1996; Turton 1999).

Societies with low or high levels of social adaptive capacity will have distinctly different abilities to deal with changing levels of water availability (Turton and Ohlsson 1999; Ashton 2002). The typical set of consequences is illustrated in Figure 8.3.

		Relative ability to develop and adopt coping strategies	
		Low	High
Relative availability of water per person	Scarce	Water poverty	Structurally-induced water abundance
	Abundant	Structurally-induced water scarcity	Water security

FIG. 8.3 A comparison of the likely outcomes of societies with two levels of 'second-order resources' (i.e. social adaptive capacity) having to deal with two levels of 'first-order resources' (i.e. water abundance or water scarcity). *Source*: Redrawn from Ashton 2002.

A society with low social adaptive capacity will be unable to deal effectively with water scarcity, thereby entering a situation called 'water poverty'. In a similar way, low levels of coping skills will prevent a society from making full use of abundant water resources, forcing it to enter a condition of 'structurally induced water scarcity'. In contrast, a high level of social adaptive capacity will allow a society to develop and implement a series of coping strategies that will permit a situation of 'structurally induced water abundance' (Turton 1999; Turton and Ohlsson 1999). A country with abundant water resources and high levels of coping skills is considered to be in a situation of 'water security' (Ashton 2002).

This straightforward comparison (Fig. 8.3) demonstrates how important it is for societies to develop, enhance, and implement appropriate sets of coping strategies if they are to deal with conditions of impending water scarcity. Inevitably, every country in southern Africa has to face the same prospect of increasing demands for water accompanied by dwindling supplies of easily accessible water that can meet these demands. Those countries that already experience 'water stress' or 'impending water deficit' (Falkenmark 1999; FAO 2000a) have to deal with this matter now. Other countries that have more water available at present will have to deal with the issue in future. In any event, an enormous effort will be needed from every country if the challenge is to be met successfully (Ashton 2002).

In contrast to South Africa (Ashton and Haasbroek 2002), and perhaps also Botswana and Namibia, other southern African countries have shown relatively low levels of 'social adaptive capacity' in their responses to the problems posed by declining water resources. In part, this appears to be due to the very different levels of social and economic development within the different countries, as reflected by each country's level of intellectual capital, infrastructure, and institutional capacity. Typically, an effective programme of national water resource management requires a country to deploy levels of social, economic, and technical resources that require considerable expenditure. Whilst no appropriate measures are available to assess and compare the efficiency and effectiveness of water resource management approaches in different southern African countries, it would appear that those southern African countries whose economies reflect greater strength and diversity, with less reliance on agriculture (Table 8.5), are likely to be in a better position to deploy appropriate institutional and technical solutions to the problems they face.

CONCLUSIONS

Every southern African country faces similar daunting pressures to stimulate national and regional development in order to alleviate poverty and improve the living standards of their populations. In particular, mining, industrial, manufacturing, and service sector activities must be expanded to provide new job opportunities, while more food will have to be grown to feed growing populations. Inevitably, all these activities will place an increased burden on the available water supplies and further complicate the management of these scarce resources. However, despite the fact that water is absolutely indispensable to all aspects of life and development, the locally available water resources in several southern African countries are insufficient to meet the projected demands that will be made of them. This crucial feature alone requires southern African countries to take a broader, concerted, and more regional approach to dealing with their needs for water (Ashton 2002).

For those countries where water shortages are a current or impending reality, water demand management offers the prospect of achieving real water savings and thereby delaying the construction of expensive water storage and supply infrastructure (Smakhtin *et al.* 2001). However, even though water demand management can improve the efficiency and effectiveness of water use, it must be seen as one component of a suite of water supply solutions that will be needed to ensure long-term water security in the region. The development of integrated strategies that span the widest possible range of technological, economic, and institutional options for water resource management hold the greatest prospect of success (Basson *et al.* 1997; Ashton 2002; Heyns 2002).

In order to improve the effectiveness and efficiency of water use patterns, more attention will have to be paid to the use of appropriate economic instruments as incentives. These can take the form of punitive measures, such as fines and levies for excessive water use or wastage, or promotional measures, such as tax incentives for improved water use. Ideally, water resource management efforts should seek to develop the most appropriate mix or combination of hard (economic and technological) and soft (social, management and institutional) solutions to suit the specific sets of circumstances. In a regional (as opposed to a national) context, this will require a high degree of alignment between national policies, legislation, strategies, and incentives of participating countries. Where these countries differ in their relative (national) levels of statutory control, technical capability, or economic and financial capacity, it will be difficult to achieve meaningful and equitable levels of success (Ashton 2002).

The central role that coping strategies should play in dealing with national and regional issues of water supply has been clearly outlined. In the water-scarce countries of southern Africa, traditional supply-driven approaches to national water resource management offer few long-term prospects of success. The development constraints posed by inadequate water supplies in specific countries can be dealt with successfully only if a wider, regional perspective is taken, in combination with concerted national (in-country) management actions. This approach will pose several challenges to the countries of southern Africa. In particular, the critical importance of good governance and the need for effective and efficient water management institutions will require each country to foster closer partnership arrangements with its neighbours. Importantly, NEPAD offers southern African countries the first real opportunity to achieve this level of harmony with their neighbours. Ultimately, the success or failure of these initiatives will depend on the legitimacy and political will of the participants.

REFERENCES

Allen, J. A., and Karshenas, M., 'Managing environmental capital: the case of water in Israel, Jordan, the West Bank and Gaza, 1947 to 1995', in J. A. Allen and J. H. Court (eds.), *Water, Peace and the Middle East: Negotiating Resources in the Jordan Basin* (London: I. B. Tauris, 1996).

Asmal, K., 'Water as a metaphor for governance: issues in the management of water resources in Africa', *Water Policy*, 1 (1998), 95–101.

Ashton, P. J., 'Water security for multi-national river basin states: the special case of the Okavango River', in M. Falkenmark and J. Lundqvist (eds.), *Proceedings of the Symposium on Water Security for Multi-National River Basin States, Opportunity for Development* (Stockholm: Stockholm International Water Institute, 2000).

—— 'Avoiding conflicts over Africa's water resources', *Ambio*, 31/3 (2002), 236–42.

Ashton, P. J., and Haasbroek, B., 'Water demand management and social adaptive capacity: a South African case study', in A. R. Turton and R. Henwood (eds.), *Hydropolitics in the Developing World: A Southern African Perspective* (Pretoria: African Water Issues Research Unit (AWIRU) and International Water Management Institute (IWMI), 2002).

——and Ramasar, V., 'Water and HIV/AIDS: some strategic considerations for southern Africa', in A.R. Turton and R. Henwood (eds.), *Hydropolitics in the Developing World: A Southern African Perspective* (Pretoria: African Water Issues Research Unit (AWIRU) and International Water Management Institute (IWMI), 2002).

Basson, M. S., van Niekerk, P. H., and van Rooyen, J. A., *Overview of Water Resources Availability and Utilisation in South Africa* (Pretoria: Department of Water Affairs & Forestry and BKS (Pty) Ltd., 1997).

Biswas, A. K., *Management of International Water: Problems and Perspective* (Paris: UNESCO, 1993).

Christie, F., and Hanlon, J., *African Issues: Mozambique and the Great Flood of 2000* (Bloomington: The International African Institute, Indiana University Press, 2001).

CIA (Central Intelligence Agency), *The World Factbook (Country Listing)*, 2000, Central Intelligence Agency of the United States of America [Online]. Available at website: http://www.cia.gov/cia/publications/factbook/geos/.

Conley, A. H., 'A synoptic view of water resources in southern Africa', in *Proceedings of the Conference of the Southern Africa Foundation for Economic Research on Integrated Development of Regional Water Resources* (Nyanga: Southern African Foundation for Economic Research, 1995).

Eberhard, R., *Supply Pricing of Urban Water in South Africa* (Pretoria: Water Research Commission, 1999).

Falkenmark, M., 'Fresh water—time for a modified approach', *Ambio*, 15/4 (1986), 192–200.

—— 'The massive water scarcity now threatening Africa: why isn't it being addressed?' *Ambio*, 18/2 (1989), 112–18.

—— 'The dangerous spiral: near-future risks for water-related eco-conflicts', in *Proceedings of the ICRC Symposium 'Water and War: Symposium on Water in Armed Conflicts'* (Montreux: International Committee of the Red Cross, 1994).

—— 'Competing freshwater and ecological services in the river basin perspective— an expanded conceptual framework', in *Proceedings of the SIWI/IWRA Seminar 'Towards Upstream/Downstream Hydrosolidarity'* (Stockholm: Swedish International Water Institute, 1999).

FAO (Food and Agriculture Organization), *New Dimensions in Water Security— Water, Society and Ecosystem Services in the 21st Century* (Rome: Food and Agriculture Organization of the United Nations, 2000a).

——*Aquastat Information Service on Water in Agriculture and Rural Development* (Rome: Food and Agriculture Organization of the United Nations, 2000b) [Online]. Available at website: http://www.fao.org/waicent/FaoInfo/Agricult/AGL/AGLW/AQUASTAT/afric.htm.

Gleick, P. H., *The World's Water 1998–1999: Biennial Report on Freshwater Resources* (Washington, DC: Island, 1998).

—— 'The human right to water', *Water Policy*, 1 (1999), 487–503.

—— 'The changing water paradigm—a look at the twenty-first century water resources development', *Water International*, 25/1 (2000), 127–38.

Heyns, P. S. v. H., 'The Namibian perspective on regional collaboration in the joint development of international water resources', *International Journal of Water Resources Development*, 11/4 (1995), 483–91.

—— 'The interbasin transfer of water between countries within the Southern African Development Community (SADC): a developmental challenge of the future', in A. R. Turton and R. Henwood (eds.), *Hydropolitics in the Developing World: A Southern African Perspective* (Pretoria: African Water Issues Research Unit (AWIRU) and International Water Management Institute (IWMI), 2002).

Heyns, P. S. v. H., *et al.*, *Namibia's Water: A Decision-Makers' Guide* (Windhoek: Desert Research Foundation of Namibia and Department of Water Affairs, Namibia, 1998).

Hounsome, R., and Ashton, P. J., *Draft Position Paper: Sustainable Development for the Mining and Minerals Sector in Southern Africa* (Johannesburg: Mining, Minerals and Sustainable Development , 2001).

Hudson, H., 'Resource based conflict: water (in)security and its strategic implications', in H. Solomon (ed.), *Sink or Swim? Water, Resource Security and State Co-operation* (Halfway House: Institute for Security Studies, 1996).

Lundqvist, J., 'Rules and roles in water policy and management—need for clarification of rights and obligations', *Water International*, 25/2 (2000), 194–201.

MMSD (Mining, Minerals and Sustainable Development—Southern Africa), *The Report of the Regional MMSD Process* (Johannesburg: Mining, Minerals and Sustainable Development, 2001).

Newson, M., *Land, Water and Development: River Basin Systems and Their Sustainable Management* (London: Routledge, 1992).

Ohlsson, L., *Water and Security in Southern Africa* (Stockholm: Swedish International Development Agency (SIDA), 1995).

OKACOM (Okavango River Water Basin Commission), *Agreement Between the Governments of the Republic of Angola, the Republic of Botswana and the Republic of Namibia on the Establishment of a Permanent Okavango River Basin Water Commission (OKACOM)*, Signatory Document, Signed by Representatives of the Three Governments, Windhoek, Namibia, 15 September 1994.

Pallett, J., *Sharing Water In Southern Africa* (Windhoek: Desert Research Foundation of Namibia, 1997).

Rosegrant, M. W., *Water Resources in the Twenty-First Century: Challenges and Implications for Action* (Washington DC: International Food Policy Research Institute, 1997).

SADC (Southern African Development Community), *Protocol on Shared Watercourse Systems in the Southern African Development Community (SADC) Region* (Gaborone: SADC Council of Ministers, 1995).

—— SADC Review and Country Profiles. Southern African Development Community Country Profiles, 2000 [Online]. Available at: http://www.sadcreview.com/countryper cent20profilesper cent202000/.

SARDC (Southern African Research and Documentation Centre), *Water in Southern Africa* (Harare: Southern African Research and Documentation Centre, 1996).

Shela, O. N., 'Water resource management and sustainable development in southern Africa: issues for consideration in implementing the Dublin Declaration and Agenda 21 in southern Africa', in *Proceedings of the Global Water Partnership Workshop* (Windhoek: Global Water Partnership, 1996).

Smakhtin, V., *et al.*, 'Unconventional water supply options in South Africa: possible solutions or intractable problems?', *Water International*, 26/3 (2001), 314–34.

Turton, A. R., 'Water scarcity and social adaptive capacity: towards an understanding of the social dynamics of managing water scarcity in developing countries', *MEWREW Occasional Paper, 9* (London: University of London, School of Oriental and African Studies (SOAS), Water Issues Study Group, 1999).

——and Ohlsson, L., 'Water scarcity and social stability: towards a deeper understanding of the key concepts needed to manage water scarcity in developing countries', in *Proceedings of the Ninth Stockholm Water Conference* (Stockholm: 1999).

UNAIDS, *Epidemiological Fact Sheets on HIV/AIDS and Country Profiles* (Geneva: UNAIDS/WHO, 2000) [Online]. Available at website: ⟨http://www.unaids.org/countryprofiles⟩.

Van der Zaag, P., and Savenije, H. H. G., 'Towards improved management of shared river basins: lessons from the Maseru Conference', *Water Policy*, 2 (2000), 47–63.

——Seyam, I. M., and Savenije, H. H. G., 'Towards objective criteria for the equitable sharing of international water resources', in *Proceedings of the Fourth Biennial Congress of the African Division of the International Association of Hydraulic Research.* (Windhoek: International Association of Hydraulic Research 2000).

Whiteside A., and Sunter, C., *AIDS: The Challenge for South Africa* (Tafelberg: Human & Rousseau, 2000).

Williams, B. G., *et al.*, 'Patterns of infection: using age prevalence data to understand the epidemic of HIV in South Africa', *South African Journal of Science*, 96/6 (2000), 305–12.

Wolf, A. T., 'Criteria for equitable allocations: the heart of international water conflict', *Natural Resources Forum*, 23 (1999), 3–30.

World Bank, *World Development Indicators 1998* (Washington, DC: Development Data Centre, World Bank, 1998).

WCD (World Commission on Dams), *Dams and Development: A New Framework for Decision-Making* (London: Earthscan, 2000).

WCED (World Commission on Environment and Development), *Our Common Future* (Oxford: Oxford University Press, 1987).

WRI (World Resources Institute), *World Resources 2000–2001* (Washington, DC: World Resources Institute, 2000), Country profiles and Data Tables [Online]. Available at website: http://www.wri.org/wr2000.

INDEX

Note: **bold** page numbers indicate chapters

Index